P9-ELG-540

recent advances in phytochemistry

volume 22

Opportunities for Phytochemistry in Plant Biotechnology

RECENT ADVANCES IN PHYTOCHEMISTRY

Proceedings of the Phytochemical Society of North America
General Editor: Eric E. Conn, *University of California, Davis, California*

Recent Volumes in the Series

Volume 15 The Phytochemistry of Cell Recognition and Cell Surface Interactions
Proceedings of the First Joint Meeting of the Phytochemical Society of North America and the American Society of Plant Physiologists, Pullman, Washington, August, 1980

Volume 16 Cellular and Subcellular Localization in Plant Metabolism
Proceedings of the Twenty-first Annual Meeting of the Phytochemical Society of North America, Ithaca, New York, August, 1981

Volume 17 Mobilization of Reserves in Germination
Proceedings of the Twenty-second Annual Meeting of the Phytochemical Society of North America, Ottawa, Ontario, Canada, August, 1982

Volume 18 Phytochemical Adaptations to Stress
Proceedings of the Twenty-third Annual Meeting of the Phytochemical Society of North America, Tucson, Arizona, July, 1983

Volume 19 Chemically Mediated Interactions between Plants and Other Organisms
Proceedings of the Twenty-fourth Annual Meeting of the Phytochemical Society of North America, Boston, Massachusetts, July, 1984

Volume 20 The Shikimic Acid Pathway
Proceedings of the Twenty-fifth Annual Meeting of the Phytochemical Society of North America, Pacific Grove, California, June, 1985

Volume 21 The Phytochemical Effects of Environmental Compounds
Proceedings of the Twenty-sixth Annual Meeting of the Phytochemical Society of North America, College Park, Maryland, July, 1986

Volume 22 Opportunities for Phytochemistry in Plant Biotechnology
Proceedings of the Twenty-seventh Annual Meeting of the Phytochemical Society of North America, Tampa, Florida, June, 1987

---

recent advances in phytochemistry

volume 22

Opportunities for Phytochemistry in Plant Biotechnology

Edited by
Eric E. Conn
University of California, Davis
Davis, California

PLENUM PRESS • NEW YORK AND LONDON

ISBN 0-306-42936-5

Proceedings of the Twenty-seventh Annual Meeting of the Phytochemical Society
of North America on Opportunities for Phytochemistry in Plant Biotechnology,
held June 21–26, 1987, in Tampa, Florida

A Division of Plenum Publishing Corporation
233 Spring Street, New York, N.Y. 10013

Printed in the United States of America

In Memory of

TSUNE KOSUGE

November 28, 1925 — March 13, 1988

PREFACE

This volume is dedicated to Tsune Kosuge in recognition of his distinguished career as a plant biochemist and his many contributions to the field of phytochemistry. Those contributions began over thirty years ago during his doctoral research at Berkeley when Professor Kosuge was examining the metabolism of coumarin precursors in leaves of *Melilotus alba*. The several papers resulting from that doctoral thesis were among the first enzymatic studies ever to be performed in the field of natural (secondary) plant products. It should also be noted that during his doctoral research Professor Kosuge obtained the first experimental evidence for the existence of phenylalanine ammonia lyase (PAL), the enzyme that controls the flow of carbon into phenylpropanoid metabolism.

After obtaining his Ph.D., Professor Kosuge returned to the discipline of plant pathology where he had obtained an M.S. and began to utilize his skills as a biochemist to examine the molecular basis of plant-pathogen interactions. A limitation on space prevents a listing of all that he has accomplished in this area, but one discovery should be mentioned, namely the enzyme in *Pseudomonas savastanoi* which catalyzes the monooxygenation-decarboxylation of tryptophan forming indole-3-acetamide. This discovery 20 years ago not only was the first description of this class of enzymes; it also laid the ground work for many papers on 3-indoleacetic acid (IAA) as a virulence factor in the interaction of *P. savastanoi* with its host plants oleander and olive. When sequencing of the genes involved in IAA formation from tryptophan by *P. savastanoi* showed a significant homology with crown gall T-DNA, the attention of numerous laboratories was drawn to Professor Kosuge's publications in this area.

Professor Kosuge has generously given of his time and energy to his department and his institution, to editorial boards of professional societies, and advisory panels of the federal government. In recent years he has become a highly respected spokesman for plant biotechnology both within the University of California and at the national level. His service at the Competitive Research Grants Office of the USDA, first in 1978-79 and later in 1983-84, was warmly welcomed and appreciated by all who interacted with that office during his tenure.

When the Phytochemical Society of North America decided to hold a symposium on the role of phytochemistry in plant biotechnology, the Executive Committee asked Professor Kosuge to serve as the organizer. We did so because of his broad knowledge of people working in the field, the problems being pursued and the progress being made. With characteristic enthusiasm Professor Kosuge was able to arrange for contributions at the symposium from eleven laboratories, several of which are among the most active in the field. The symposium which was held on June 22-27, 1987 at the University of South Florida in Tampa, provided for an unusual degree of interaction between the speakers, some of whom were only a few years older than many of the graduate students and post-doctorals in the audience.

The nine chapters resulting from the symposium are arranged into three groupings. Chapters 1 through 4 describe specific strategies for studying problems in plant biotechnology; the strategies range from studies with whole plants to work on the primary structure of an enzyme that serves as the target for herbicide action of glyphosate. Two chapters (chapters 5 and 6) describe studies on the role of specific phytochemicals in important physiological processes. Chapters 7 and 8 deal with the general subject of plant stress, either from disease or from insect attack, and the role of phytochemistry in these model systems. Chapter 9 is an amusingly candid discussion, by an experienced observer of the field, of the problems of justifying and funding plant biotechnology in an era of excess farm products and agricultural self-sufficiency.

The Organizing Committee for the 27th Annual Meeting of the Phytochemical Society of North America, at which this symposium was presented, consisted of Richard L. Mansell and John T. Romeo who served as gracious hosts. As noted above, the PSNA is grateful to Tsune Kosuge for his choice of speakers and topics for the symposium. The facilities of the University of South Florida provided an excellent environment for the meeting.

Financial support was provided by the College of Natural Sciences and the Division of Sponsored Research of the University of South Florida, the Dionex Corporation, E.I. duPont de Nemours and Company, the Monsanto Company, and the Phytochemical Society of North America. Thanks are also due to Ms. Billie Gabriel for her skillful preparation of the camera-ready copy.

Eric E. Conn

February 1988

CONTENTS

Chapter One

SOMACLONAL VARIATION: ITS GENETIC BASIS AND PROSPECTS FOR CROP IMPROVEMENT

ROBERT A. MORRISON, R.J. WHITAKER, AND DAVID A. EVANS

DNA Plant Technology Corporation
2611 Branch Pike
Cinnaminson, New Jersey 08077

INTRODUCTION

Somaclonal variation is defined as genetic variation observed among progeny of plants regenerated from somatic cells cultured *in vitro*. Although theoretically all plants regenerated from somatic cells should be clones, a number of observations have indicated that this is not the case.[1,2] In addition to the basic genetic implications of this phenomenon, the variation has proven useful in breeding programs of various crop plants.[3] Conventional plant breeding has traditionally been the sole avenue for crop improvement; however, recent demands of agricultural-based industry has shown that this process is too time consuming for typical corporate time frames. While variability has been obtained from wild species of cultivated crops its access has been quite limited. Indeed many desired traits such as disease resistance have yet to be identified. It is also difficult to establish novelty related to legal protection of conventional breeding outputs since many workers are using the same breeding lines in their programs. Thus somaclonal variation provides great promise for reducing the time required to produce new varieties or breeding lines which are easily patentable due to their novel variation.

GENETIC BASIS OF SOMACLONAL VARIATION

While the genetic basis of somaclonal variation was not determined in early reports of this phenomenon, more recent work has included genetic analysis of the variation.[3] The consensus from this work has been that the origin of somaclonal variation is both the inherent variation in the tissue which is placed in culture and changes which are associated with the passage through tissue culture.[3,4] Results with wheat indicate that the genome of higher plants undergoes constant change, and the regeneration of plants from single cells effectively captures this variation. For instance, genetic changes in somatic cells of stems and leaves have been documented.[5] Phenomena such as gene amplification or chromosome rearrangements appear to be related to the function of the tissue; yet the continuity of the genome is maintained through inheritance since the cells harboring these changes do not contribute to the germline. However, if these variant cells are induced to become plants, then the variants would be observed in the regenerated plants as well as their progeny. Variation among regenerated plants may also be associated with the passage through tissue culture.[3] Much of this variation may be related to substances in the culture media that act as mutagens or maintain a degree of cell division for which the plant cell's repair mechanisms cannot keep pace. An example of a substance in the culture medium behaving as a mutagen was described by Evans and Bravo among regenerated plants of ornamental tobacco.[6] They observed that an increase in the level of the cytokinin 6-benzyladenine (6BA) resulted in a greater number of variant plants. Plants regenerated on 5-10 μM 6BA were often shorter compared to plants regenerated in the presence of lower concentrations of the growth regulator. In addition, a trend of decreased pollen viability among regenerated plants derived from high concentrations of 6BA was observed.

Growth regulators may also be the indirect basis for quantitative and structural chromosome changes. Auxins and cytokinins, two of the most common plant growth regulators, affect cell division and thus indirectly may

Table 1. Summary of Types of Altered Traits Resulting From Somaclonal Variation in Various Crop Plants.

Crop	Trait	Reference
Rapeseed	Seed color	7
Tomato	Fruit color	8
Tomato	Disease resistance	2
Alfalfa	Flower color	9
Tobacco	Leaf color	10
Tobacco	Leaf spots	11
Potato	rDNA Copy number	12
Potato	mt DNA	12
Maize	Adh Isozyme	13

contribute to gross chromosomal changes by forcing the cell into the next phase of the cell cycle before it has a chance to repair mistakes. Increase in ploidy could clearly result from a disruption of the cell cycle by one of these growth regulators. A summary of several crops for which somaclonal variation has been documented is presented in Table 1.

While variation among plants regenerated from tissue culture has been noted by many authors, the variation cannot be attributed as genetic unless it is observed among the progeny of the regenerated plant.[14] A number of variants, both at the cell and whole plant level, have not been observed among progeny of regenerated plants.[15] Non-heritable variation is termed epigenetic variation and probably results from an effect of the *in vitro* phase on the expression of a gene(s). For example, a fruit-specific gene conferring yellow color may be turned on as a result of the passage through tissue culture; however, when gametes are formed, the defect is repaired and the progeny are normal.

Mitotic crossing over (MCO) could also account for some of the variation detected in regenerated plants. This could include both symmetric and asymmetric recombination. MCO may account for the recovery of homozygous recessive single gene mutations in some regenerated plants (cf. Reference 8). As breeders have previously only had access to variation that is normally transmitted through meiosis, the recovery of products of MCO may constitute a unique source of new genetic variation.

Despite the lack of genetic data, several authors have also speculated that transposable elements may also be responsible for somaclonal variation. Variation in the insertion of plasmid-like DNA found in mitrochondria of cms-s corn has been detected in corn cell cultures.[16] Heterozygous light-green (Su/su) somaclones with a high frequency of colored spots of the leaf surface have been detected for a clone of both *Nicotiana tabacum*[17] and a *N. tabacum* + *Nicotiana sylvestris* somatic hybrid.[18] The somatic hybrid has an unstable pattern of inheritance that would be consistent with an unstable gene, although detailed genetic analysis has not yet been completed.

One of the best characterized cases of somaclonal variation has been tomato.[3] The variation was observed among progeny of plants regenerated from leaf tissue of a standard open-pollinated variety of tomato. Seeds of progeny were sown in the greenhouse and young plants were transferred to the field in replicated plots. Detailed notes were collected in the greenhouse and the field. Chromosomal variants, single gene changes, and cytoplasmic genetic variants have all been detected among the progeny of somaclones.

Chromosomal variants were detected, particularly tetraploids, $2n = 48$, at a frequency of approximately ten percent. In addition, 13 single gene mutations were observed among a total of 230 regenerated plants in one experiment. Mutations were observed for traits ranging from leaf and fruit color to jointless pedicels.[8] In the case of fruit color, an individual regenerated plant produced red fruit; however, several progeny plants had yellow fruit. Although the majority of the progeny exhibited red fruit,

Table 2. Progeny and Mapping Analysis of Yellow-Fruited Somaclonal Variant.

Progeny	Red	Yellow
R_0	15	0
R_1	15	4
R_2 From yellow R	0	99
R_2 From homozygous red R	132	0
R_2 From Heterozygous Red R	227	68
Yellow R_2 x r-2/r-2	0	24
Yellow R_2 x t/t	15	0

the observation of yellow-fruited progeny indicated that a mutation had occurred that was recessive to red fruit color. In that tomato is a crop for which extensive genetic resources are available, known yellow-fruited mutants were crossed with the somaclonal mutant in order to identify its location in the tomato genome.

When crossed with a line homozygous for the recessive mutation that confers tangerine colored fruit, all hybrid progeny were red-fruited indicating complementation. However, when the yellow-fruited somaclonal variant was crossed to a line homozygous for a recessive mutation causing yellow fruit, hybrid progeny were yellow demonstrating that these mutations resided at the same locus on Chromosome 3. The progeny and mapping analysis are presented in Table 2.

Similar strategy has been used to map other mutations observed among regenerated tomato plants including orange fruit which mapped to the distal end of Chromosome 10. Other mutations conferring jointless pedicels and resistance to the fungal pathogen *Fusarium oxysporum* reside at opposite ends of Chromosome 11.

Although many variants observed among regenerated plants have been genetically characterized, other observed variants indicate that unique variation has occurred. Some of this unique variation includes changes in regulatory genes and organelle DNA, and also mitotic crossing over (MCO). MCO was implicated as a genetic mechanism for somaclonal variation by using a tomato line heterozygous at four marked loci on Chromosome 6. From a total of 61 regenerated plants, 42 were parental types while 19 exhibited recombination for one or more of the markers.

Another mechanism that appears to be contributing to somaclonal variation is change in organelle DNA. Analyses of mitochondrial DNA of plants regenerated from maize plants carrying T cytoplasm with characteristics of N cytoplasm revealed sequence differences among the regenerated plant.[19] In addition, Kemble and Shepard (1984) detected changes in mitochondria but not chloroplast DNA of potato plants regenerated from isolated protoplasts.[20] The origin of these cytoplasmic changes may occur during the callus phase just before shoot regeneration when the number of organelles and the DNA copy number within organelles is greatly reduced.[21] If the change occurs when the cells are undergoing rapid reproduction, it is likely to become stably inherited.

Another form of novel variation observed among regenerated plants of tomato involves changes in regulatory genes. These types of genes are encoded in the nucleus and affect diverse biochemical pathways. Somaclonal variation for one such gene was observed among the progeny of a regenerated plant. Several of the progeny exhibited mottled leaves characterized by green and white sectors in the leaf. Although mutations that affect pigmentation within leaves do not necessarily affect the fruit, this particular mutation affected both as evidenced by the appearance of mottling in the fruit. In contrast to this variant, a second mutant that exhibited virescence in the leaves was associated with tangerine-colored fruit. Genetic analysis of this variant revealed that it was the result of a single recessive mutation. Analysis of pigment content in fruits and leaves of these two variants demonstrated that in mottled plants, lycopene and chlorophyll levels were affected while tangerine virescent plants had abnormal lycopene levels. However, the tangerine virescent mutation was expressed only in young leaves as seen from chlorophyll assays in the various tissues. Thus, with respect to these two variants, it appears that somaclonal variation is a random process affecting not only genes involved in morphological aspects but also regulatory genes that may control the expression of other genes. Indeed the tangerine virescent variant may involve a mutation in a gene which regulates the amount of chlorophyll in young leaves but not older leaves.

GAMETOCLONAL VARIATION

As described above, the types of genetic changes that are recovered in plants regenerated from cell culture are dependent upon the donor material that is used for plant regeneration. For tomato we used a diploid inbred variety; hence, most of the variant regenerated plants were heterozygous resulting in segregation of new mutations in the R_1 generation. Lorz and Scowcroft used heterozygous material so that mutations could be visually detected in regenerated plants.[17] Among these somaclones several plants had distorted segregation ratios in the R_1 generation. Since the generation of variation will ultimately be used for breeding and crop improvement, it is important to distinguish somatic-derived mitosis from meiosis, governed by Mendel's laws of segregation and independent assortment. Three genetic differences should be pointed out as evidence that gametoclonal and somaclonal variation are distinct. (1) Both dominant and recessive mutants induced by gametoclonal variation will be expressed directly in haploid regenerated plants from diploid anthers, since only a single copy of each gene is present. Hence, regenerated gametoclones (R_0) can be analyzed directly to identify new variants. (2) Recombinational events that are recovered in gametoclones would be the result of meiotic crossing over, not mitotic crossing over. While not fully characterized, evidence from *Neurospora* suggests that these two phenomena do not occur at the same frequency along the gene map. For example, MCO may be used to separate genes that are hard to separate by meiotic crossing over. (3) To use gametoclones, chromosome number must be doubled. The most frequently used method to double chromosomes is treatment with colchicine. Colchicine is known to induce mutations.[22] Hence, by the time gametoclones are genetically analyzed, some of the observed variation may not be due to gametoclonal variation but may be due to mutagenic effects of colchicine. Gametoclonal variation has been reported for a number of crop plants,[23] however, like somaclonal variation, the limiting factor has been the inability to regenerate plants from crop plants at will.

Gametoclonal variation was detected among doubled-haploids (DHs) of the bell pepper (*Capsicum annuum*) variety Emerald Giant.[24] The majority of the DHs were significantly shorter than the control lines; however, several were observed to be significantly taller. Variation was also observed for yield and dry weight of fruit. While deleterious variants were observed, several DH lines out-performed the control lines for yield and fruit quality.

APPLICATION OF SOMACLONAL VARIATION

In that somaclonal variation has proven to produce genetic changes in crop plants, it should be possible to exploit this variation for varietal improvement. Since single gene and organelle mutations have been documented among somaclonal variants, it should be possible to make incremental improvements in the best available varieties. Thus somaclonal variation can be used to obtain new variants that retain all the favorable characteristics of the existing variety while adding one or two additional traits such as disease resistance, herbicide resistance, or yield improvement. In practice this would involve tissue culture designed to regenerate plants from non-meristemic explants on a medium that induces variation at a high frequency. Once R_1 lines are obtained through selfing regenerated plants, they are evaluated in the field and improved lines or single plants within a line are selected based upon the specific criteria for which improvement is required. Progeny of selected plants or lines are then field tested in replicated plots in several locations to verify the stability of the new trait. Lines that continue to perform well can be used for additional rounds of somaclonal variation for continued incremental improvement of the variety for maximum performance. Because the regeneration of plants and subsequent field evaluation can be accomplished within two years, it is possible to produce new breeding lines with desirable traits in a short period of time.

Somaclonal variation is a random process and thus improvement of any given trait involves regeneration of a large number of lines and subsequent field evaluation of their progeny. Extremely deleterious variants, however, are excluded during the process by two important selection steps that serve as sieves to permit recovery of a population of R_1 plants that are most suitable for a breeding program. During the tissue culture phase, only cells that have a competent genome in terms of genes involved in morphogenesis will be able to regenerate since mutation in these genes would slow or inhibit regeneration. The second level of selection occurs in the greenhouse during flower development, fertilization, and fruit and seed set. If the plant is genetically deficient for any of these characters, seed will not be obtained. Hence, R_0 plants with deleterious changes in morphogenesis are eliminated and the lines generated for evaluation are most suitable for rapid variety development.

Somaclonal variation technology has been applied for improvement of tomato varieties for characteristics such as increased soluble solids and disease resistance. One somaclonal variant was identified with a 20-percent increase in fruit soluble solids over the processing tomato variety UC82B from which it was derived. Other somaclone derivatives of this variety have been evaluated for other characteristics including yield and pigmentation resulting in improvements for both of these traits.

An important application of somaclonal variation technology in variety improvement is incorporation of disease resistance into commercial cultivars. Although conventional plant breeding techniques have proven successful in transferring genes conferring resistance to various plant pathogens, the ability to make changes in only a few genes indicates the usefulness of somaclonal variation for development of disease resistant varieties. An example of this strategy is a UC82B somaclone derivative (DNAP 17) that was found to be resistant to the fungal pathogen *Fusarium oxysporum* Race 2. Some progeny of a regenerated UC82B somaclone were resistant to pathogen inoculated on roots. Resistant progeny were observed at a frequency of 3 resistant: 1 sensitive, indicating that somaclonal variation had induced a single gene mutation for this trait. Subsequent genetic analysis involving crosses with cultivars that were either sensitive or resistant to the pathogen revealed the resistance in DNAP 17 was indeed conferred by a single gene. In addition, the analysis demonstrated that the mutation resided at the same locus as the resistance obtained by conventional methods. Since complementation was not observed, it was concluded that the resistance obtained by somaclonal variation was identical to the resistance trait derived from conventional breeding.

SECONDARY METABOLISM

In addition to aspects of crop improvement involving agronomic characteristics such as yield and disease resistance, somaclonal variation should prove beneficial for increasing the production of plant-derived chemicals. This aspect of plant biotechnology can be used to boost the chemical content in whole plants or cell cultures by exploiting the random variation occurring in specialized cell cultures used to produce somaclonal variants. The key to technical and economic feasibility rests on the ability to induce and select genetically stable whole plants or cell cultures that overproduce specific chemicals and the development of scale-up technology that exploits the biological capabilities of plant cells and promotes efficient chemical production.

The choice between whole plant or cell culture production is dependent upon several factors including value of the compound, market size, chemical complexity, technical feasibility, and the amount of time allowed to improve the chemical source. Hence, low-cost chemicals (less than $500 per kg) would be better suited for whole plant or agricultural production while more expensive, plant-derived chemicals could be produced from cell cultures similar to what is now possible using microbial fermentation systems.

Agricultural production of chemicals could be directly improved by somaclonal variation using a strategy similar to that described for other agronomic traits coupled with an assay to detect increased levels of chemical accumulation. Candidates for this approach include pyrethrins (potent natural insecticides produced in the florets of the *Pyrethrum* daisy), natural sweeteners with high sucrose potency equivalents, special composition edible and industrial oils (high oleic acid oils or oils with increased levels of Omega-3 fatty acids), flavors, and fragrances.

Modern agricultural innovations offer potential advantages over current methods of sourcing plant chemicals; systematic fertilization regimes, and cultivation methods would surely increase plant vigor and thus chemical production. In that breeding programs and introgression of variation from wild species has proven beneficial for improvement of virtually all crops, similar strategies would have significant impact upon chemical yields. Thus, the shortened time frame provided by the production of somaclones in achieving plant breeding objectives could result in cultivars with increased biosynthetic capacities that could undergo large-scale production as an additional means of boosting overall productivity. Somaclones could also be tested in hybrid form to further increase the production levels afforded by hybrid vigor. As previously stated, from an economic perspective, agricultural chemical production is a more cost-efficient and workable proposition than cell culture production for low-cost products. Agricultural production, augmented by a biotechnology-driven crop improvement program, can be fully operational at commercial scale in a 2-4 year time frame while technical developments are pursued for tissue culture production.

Chemical production in cell cultures is dependent upon technical achievements in scale-up and bioreactor development. These achievements will narrow the cost gap and time factor between whole plant and cell culture production especially for higher value chemicals. Although agricultural production is attractive due to practicality and feasibility, in many instances bioreactor production of secondary metabolites may be more desirable for logistic and proprietary reasons. Agricultural chemical production is subject to uncontrollable environmental factors such as drought, heat, cold, and pests that can affect the quantity and quality of the desired chemical. In addition, the seasonal nature of agriculture can prohibit securing year-round supplies of chemicals that must be fresh. Hence, bioreactor production of valuable chemicals would afford a greater degree of control and flexibility.

Assuming technical feasibility of chemical production in bioreactors, cell lines capable of producing chemicals *in vitro* would then be required. Economic feasibility ultimately rests in the ability to obtain genetic variants that overproduce desirable chemicals. Thus somaclonal variation would be useful to obtain such variants by affecting changes in genes related to the desired biosynthetic pathway or in genes involved in excretion of the product from the cell. A program designed to generate high-producing cell cultures must begin with the development of a sensitive assay procedure that permits the screening of large numbers of samples. Also, medium optimization must be accomplished to insure that aspects associated with _in vitro_ growth do not become the limiting factors for successful production of chemicals. An example of the importance of media composition and its impact on the technical and economic feasibility of tissue culture production of chemicals is the optimization of shikonin synthesis in cell cultures of _Lithosperum erythrorhizon_. This red pigment which also possesses pharmaceutical properties was the first plant secondary metabolite produced by cell cultures on a commerical scale. Success was dependent upon two distinct media formulations. One permitted rapid cell proliferation while the second promoted shikonin biosynthesis. Specific components of the production medium were found to be critical for optimal production including copper, nitrate, sulfate, and sucrose concentration. Based upon these findings regarding economic considerations, somaclonal variation coupled with an accurate assay should prove useful for producing cell lines that could produce a product at optimal levels without the addition of costly ingredients required for shikonin production.

FUTURE APPLICATIONS

Somaclonal variation observed among regenerated plants of several crops has stimulated interest in the application of this technology in plant breeding programs. In that somaclonal variation affords the ability to make discrete genetic changes that are stably inherited, its importance in crop improvement will continue to grow. Due to the shortened time frame and access to novel variability, application of somaclonal variation as a breeding tool will become more commonplace.

Progress in tomato improvement by somaclonal variation has proven that this technology should be beneficial for improvement of other crops. It appears that virtually any characteristic that can be manipulated through conventional inbreeding methods should likewise be improved by somaclonal variation. An added potential benefit of somaclonal variation that may further decrease the time frame of breeding line or variety release is cellular selection.

Although selection *in vitro* of plants exhibiting increased yield or soluble solids has not been realized, a better understanding of the genetic mechanism of such traits may one day result in a simplified assay that can be coupled to a somaclonal variation system. More realistic is the combination of this technology to selection of plants exhibiting tolerance to salt or heavy metals or resistance to herbicides or pathogen toxins. Feasibility of this strategy rests on the correlation between the cellular and whole plant response to specific substances used as selective agents. For instance plants regenerated from cells that are capable of growing in high salt concentrations do not necessarily exhibit salt tolerance. It seems that a genetic change involving permeability of a callus cell to salt may not be expressed in cells of the roots. The random nature of somaclonal variation can be exploited by regenerating a large number of plants from various explants that have been exposed to high levels of salt and the progeny evaluated for salt tolerance. Due to the possibility of recessive traits conferring tolerance or resistance to the selective agent, haploid tissue can be used as a source of explants thereby allowing expression of recessive traits in the R_0 plant.

Promising results have already been obtained by selecting for resistance to host-specific pathogen toxins, herbicide resistance, and resistance to heavy metals. A number of substances have been incorporated into plant cell culture media with the objective of isolating cell lines capable of growth and ultimately demonstrating inheritance of the resistance in regenerated plants.[25] However, many reports have used plant species that are difficult to regenerate. Other workers employed selective agents for which resistance is difficult or impossible to assay in whole plants.

Chlorosulfuron was used as a selective agent in tobacco cell cultures to develop plants that were resistant to this herbicide.[26] In that this substance affects a basic cell function, it was reasoned that resistance exhibited by callus cells would be expressed at the whole plant level. Regeneration of plants from haploid tobacco callus that grew in the presence of the herbicide was accomplished. Subsequence genetic analysis revealed that resistance was observed among the progeny in a ratio expected for a single semi-dominant gene.

In vitro selection for disease resistance appears to be an attractive application of somaclonal variation provided a phytotoxin is available as a selective agent. Ling *et al.*[27] identified rice plants resistant to brown spot disease from *in vitro* selection using phytotoxin from the rice pathogen *Helminthosporium oryzae*. Rice calli shaken in a medium containing various concentrations of toxin that continued to grow were regenerated into plants, and their progeny evaluated for disease resistance. While detailed genetic analysis was not performed they reported segregation ratios indicating that the resistance was conferred by two dominant genes.

Of special note in this study regarding somaclonal variation is the regeneration of a resistant plant from the toxin-free control series. A similar result was reported by Shahin and Spivey[28] using protoplasts of tomato grown in the presence of the non-specific toxin fusaric acid derived from the *F. oxysporum*. Using the tomato variety UC82B which is susceptible to *Fusarium* wilt, plants were regenerated from protoplasts grown on a medium supplemented with fusaric acid as well as toxin-free media. Progeny analysis of regenerated plants revealed resistance to the pathogen among plants derived from selection with the toxin as well as among plants regenerated from the toxin-free control medium.

Although in both studies all plants derived from seeds of the starting material were susceptible, the observation of a resistant plant among non-selected variants points to the promise of developing resistant varieties of plants to other pathogens through somaclonal variation. Although the incorporation of these variants into a breeding program has not been reported, similar work with tomato indicates the feasibility of varietal development using this technology.[8]

INTERFACE WITH MOLECULAR BIOLOGY

In addition to the direct use of somaclonal variation as a breeding tool, other more basic applications are possible. The field of recombinant DNA technology has been reported to have great potential for crop improvement by providing transfer of single gene traits in a relatively short period of time. However, the only effective means of transforming higher plants with recombinant genes involves the use of the Ti plasmid vector harbored by *Agrobacterium tumafaciens*. In addition selected transformed cells (leaf disc, protoplasts) must undergo regeneration *in vitro* which may result in variation among the resulting plants. The most efficient methods require plant regeneration from transformed cells or protoplasts. Unfortunately, this requirement demonstrates a deleterious aspect of somaclonal variation since it has been noted that following transformation of tomato with *Agrobacterium*, cells with a wide array of chromosome numerical and structural variation were detected. Since this level of chromosome variation was not observed in cell cultures from which desirable variants were obtained, transformation coupled with a systematic somaclonal variation procedure should reduce the amount of deleterious variants. This will ensure that the goal of single gene transfers, desired of transformation, will be realized.

Thus the study of somaclonal variation in individual crops will undoubtedly provide valuable information to minimize variation in specialized tissue culture systems such as *Agrobacterium* cocultivation. Manipulation of media components would provide the most rapid route from protoplast to plant with minimal duration of the callus phase. Studies regarding correlation between specific tissues and the amount of variation observed among regenerated plants would identify appropriate explant sources for cocultivation. Somaclonal variation may be applied directly to higher plant transformation in that somaclones may be identified that are more suitable for uptake, integration, and stable expression of foreign DNA.

As somaclones are derived from the plant regeneration process, they may prove more amenable to transformation procedures requiring plant regeneration. Also, it may be possible to identify somaclonal variants that are more efficient for uptake and/or integration of DNA. Variants may be identified that exhibit increased expression of recombinant genes derived from transformation. The potential for the benefit of somaclonal variation in transgenic plants was reported by Peerbolte *et al.*[29] These workers observed rearrangements in T-DNA of transgenic tobacco cell lines resulting in inactivation of the non-rooting and octopine synthesis functions allowing shoot formation. This report sheds light on the potential for using somaclonal variation to select for plants with integration of T-DNA into specific regions of the plant genome.

Finally, as somaclonal variation results in a large number of simple genetic changes, many somaclones are isogenic stocks of the original parental variety. In that these somaclones differ from the parent by only a single or small number of genetic changes, somaclones could be a valuable source of germplasm to isolate genes for agronomically useful characters. Comparison of mRNA patterns of the parent and a somaclone exhibiting improvement in a specific character would result in distinct mRNAs for the specific phenotype. These messages could be used to isolate the genes for subsequent transformation of other breeding lines or species.

REFERENCES

1. LARKIN, P.J., W.R. SCOWCROFT. 1981. Somaclonal variation - a novel source of variability from cell cultures for plant improvement. Theor. Appl. Genet. 60: 197-214.
2. EVANS, D.A., W.R. SHARP, H.P. MEDINA-FILHO. 1984. Somaclonal and gametoclonal variation. Amer. J. Bot. 71: 759-774.
3. EVANS, D.A., W.R. SHARP. 1986. Somaclonal and gametoclonal variation. *In* Handbook of Plant Cell Culture. (D.A. Evans *et al.*, eds.), Vol. 4, Macmillan Publishing Company, New York, pp. 97-132.
4. SCOWCROFT, W.R. 1985. Somaclonal variation: the myth of clonal uniformity. *In* Genetic Flux in Plant. (B. Hahn, E.S. Dennis, eds.), Springer Verlag, New York, pp. 112-156.
5. D'AMATO, F. 1952. Polyploidy in the differentiation and function of tissues and cells in plants. A critical examination of the literature. Caryologia 4: 311-357.

6. EVANS, D.A., J.E. BRAVO. 1986. Phenotypic and genotypic stability of tissue cultured plants. In Tissue Culture as a Plant Production System for Horticultural Crops. (R.H. Zimmerman et al., eds.), Martinus Nijhoff Publishers, Dordrecht, The Netherlands, pp. 73-94.
7. GEORGE, L., P.S. RAO. 1983. Yellow seeded variants in in vitro regenerants of mustard (Brassica juncea cross var. RA1-5). Plant Sci. Lett. 30: 327-330.
8. EVANS, D.A., W.R. SHARP. 1983. Single gene mutations in tomato plants regenerated from tissue culture. Science 221: 949-951.
9. GROOSE, R.W., E.T. BINGHAM. 1984. Variation in plants regenerated from tissue culture of tetraploid alfalfa heterozygous for several traits. Crop Sci. 24: 655-658.
10. DULIEU, H., M. BARBIER. 1982. High frequencies of genetic variant plants regenerated from cotyledons of tobacco. In Variability in Plants Regenerated from Tissue Culture. (L. Earle, Y. Demarly, eds.), Praegar Press, New York, pp. 211-299.
11. THANUTONG, P., I. FURUSAWA, M. YAMMATO. 1983. Resistant tobacco plants from protoplast-derived calluses selected for their resistance to Pseudomonas and Alternaria toxins. Theor. Appl. Genet. 66: 209-215.
12. LANDSMANN, J., H. UHRIG. 1985. Somaclonal variation in Solanum tuberosum detected at the molecular level. Theor. Appl. Genet. 71: 500-505.
13. BRETTELL, R.I.S., E.S. DENNIS, W.R. SCOWCROFT, W.J. PEACOCK. 1986. Molecular analysis of a somaclonal variant of maize alcohol dehydrogenase. Mol. Gen. Genet. 202: 235-239.
14. CHALEFF, R.S. 1981. Genetics of higher plants. Applications of cell culture. Cambridge University Press, New York.
15. MEINS, F. 1983. Heritable variation in plant cell culture. Ann. Rev. Plant Physiol. 34: 327-346.
16. CHOUREY, P.S., R.J. KEMBLE. 1982. Transposition event in tissue cultured cells of S-cms genotype of maize. Maize Genet. Coop. Newslett. 56: 70.
17. LORZ, H., W.R. SCOWCROFT. 1983. Variability among Plants and their progeny regenerated from protoplasts of Su/su heterozygotes of Nicotiana tabacum. Theor. Appl. Genet. 66: 67-75.

18. EVANS, D.A., J.E. BRAVO, S.A. KUT, C.E. FLICK. 1983. Genetic behavior of somatic hybrids in the genus _Nicotiana_: _N_. _otophora_ + _N_. _tabacum_ and _N_. _sylvestris_ + _N_. _tabacum_. Theor. Appl. Genet. 65: 93-101.
19. KEMBLE, R.J., R.I.S. BRETTELL, R.B. FLAVELL. 1982. Mitochondrial DNA analyses of fertile and sterile maize plants from tissue culture with the Texas male sterile cytoplasm. Theor. Appl. Genet. 62: 213-217.
20. KEMBLE, R.J., J.F. SHEPARD. 1984. Cytoplasmic DNA variation in a potato protoclonal population. Theor. Appl. Genet. 69: 211-216.
21. BENDICH, A.J., L.P. GAURILOFF. 1984. Morphometric analysis of cucurbit mitochondria: the relationship between chondriome volume and DNA content. Protoplasma 119: 1-7.
22. SANDERS, M.E., C.J. FRANZKE. 1976. Effect of temperature on origin of colchicine-induced complex mutants in sorghum. J. Hered. 67: 19-29.
23. MORRISON, R.A., D.A. EVANS. 1987. Gametoclonal variation. _In_ Plant Breeding Reviews. (J. Janick, ed.), Vol. 5, AVI Publishing, Westport, Connecticut, pp. 359-391.
24. MORRISON, R.A. 1986. Gametoclonal Variation in Pepper. Ph.D. Dissertation, Rutgers - The State University of New Jersey, Piscataway, New Jersey.
25. FLICK, C.E. 1984. Isolation of mutants from cell cultures. _In_ D.A. Evans _et al_., eds., _op_. _cit_. Reference 3, Vol. 1, pp. 393-441.
26. CHALEFF, R.S. 1983. Isolation of agronomically useful mutants from plant cell cultures. Science 219: 676-682.
27. LING, D.H., P. VIDHYASEHARAN, E.S. BORROMEO, F.J. ZAPATA, T.W. MEW. 1985. _In vitro_ screening of rice germplasm for resistance to brown spot disease using phytotoxin. Theor. Appl. Genet. 71: 133-135.
28. SHAHIN, E.A., R. SPIVEY. 1986. A single dominant gene for _Fusarium_ wilt resistance in protoplast-derived tomato plants. Theor. Appl. Genet. 73: 164-169.
29. PEERBOLTE, R., P. RUIGROK, G. WULLEMS, R. SCHILPEROORT. 1987. T-DNA rearrangements due to tissue culture: somaclonal variation in crown gall tissues. Plant Molec. Biol. 9: 51-57.

Chapter Two

GENETIC MANIPULATION OF THE FATTY ACID COMPOSITION OF PLANT LIPIDS

CHRIS R. SOMERVILLE AND JOHN BROWSE

*MSU-DOE Plant Research Laboratory
Michigan State University
East Lansing, Michigan 48824

+Institute of Biological Chemistry
Washington State University
Pullman, Washington 99164

INTRODUCTION

There are, in principle, many attractive opportunities for using recombinant DNA techniques to manipulate the lipid metabolism of higher plants. For instance, there are currently no major field crops that are used as a source of medium chain fatty acids (C8 - C12). If we understood the factors that regulate the acyl group chain length of storage lipids, it might be possible to genetically engineer one or more crop species to produce medium or very long chain fatty acids. Similarly, if detailed information were available concerning the enzymes which regulate fatty acid desaturation, it might be possible to use cloned genes for these enzymes to manipulate the fatty acid composition of many species to suit specific industrial needs. There are also many conceivable applications of recombinant DNA techniques to the manipulation of membrane lipid composition for both applied and academic ends. For instance, because of the possible importance of membrane lipid composition in the temperature responses of plants,[1] it may be possible to genetically modify economically important crop species to better suit particular environmental conditions.

Unfortunately, as in many other aspects of plant biology, the lack of detailed specific information about plant lipid biochemistry is a barrier to the application of genetic engineering techniques. The problems may be subdivided into two areas. First, there is the general problem of how the various lipid species are synthesized. This includes the related problems of where in the cell lipids are made, how the synthesis is regulated and how the composition of different membranes is controlled. Second, there is the problem of lipid function. Each membrane in the cell has a distinct lipid composition with respect to both head group and acyl composition. In general, we do not know why membranes have a particular composition or why the composition of one membrane differs from any other membrane in a cell. The same general questions pertain to deficiencies in our understanding of the biosynthesis and function of storage lipids in seeds and fruits. This latter area is of direct relevance to agricultural biotechnology because of the many potential opportunities which exist to modify the composition of storage lipids by genetic engineering techniques and the increasing demand for specialty oil products in food and manufacturing industries.

In this article we have attempted to summarize what we believe to be some of the opportunities that exist to advance our understanding of plant lipid metabolism by the use of both conventional and molecular genetics. We believe that the application of molecular genetics to problems in lipid biochemistry is useful not only as a complementary approach to existing experimental methods, but is also an important first step in the development of the specific knowledge that will be required to undertake genetic engineering of plant lipid composition.

THE TWO PATHWAYS FOR GLYCEROLIPID BIOSYNTHESIS

It is now generally accepted that there are two distinct pathways in plant cells for the biosynthesis of glycerolipids and the associated production of polyunsaturated fatty acids. The evidence for the "two pathway" model has been summarized in a review by Roughan and Slack.[2] In brief, the model proposes that fatty acids synthesized de novo in the chloroplast may either be used directly for production of chloroplast lipids by a pathway in the chloroplast (the "prokaryotic pathway"), or may be exported to the cytoplasm as esters of Coenzyme A (CoA) where they are incorporated into lipids in the endoplasmic reticulum by an independent set of acyltransferases (the "eukaryotic pathway"). The essential features of this model are depicted in Figure 1. Both pathways are initiated by the synthesis of the palmitoyl ester of acyl carrier protein (16:0-ACP) by the fatty acid synthetase in the plastid. 16:0-ACP may be elongated to 18:0-ACP and then desaturated to 18:1-ACP by a soluble desaturase so that 16:0-ACP and 18:1-ACP are the primary products of plastid fatty acid synthesis. These thioesters may be used within the chloroplast for the synthesis of phosphatidic acid (PA) by acylation of glycerol-3-phosphate, or they may be hydrolyzed to free fatty acids which move through the chloroplast envelope to be converted to thioesters of CoA in the outer envelope membrane by acyl-CoA synthetase.

Because of the substrate specificities of the plastid acyltransferases[3] the PA made by the prokaryotic pathway has a 16-carbon fatty acid at the sn-2 position and an 18-carbon fatty acid at the sn-1 position. This PA is used for the synthesis of phosphatidylglycerol (PG) or is converted to diacylglycerol (DAG) by a PA-phosphatase located in the inner chloroplast envelope.[4,5] This DAG pool can act as a precursor for the synthesis of the other major thylakoid lipids monogalactosyl diacylglycerol (MGD), digalactosyl diacylglycerol (DGD) and sulfolipid (SL).[6,7]

Acyl groups exported from the chloroplast as CoA esters are used for the synthesis of PA, mainly in the endoplasmic reticulum. In contrast to the plastid isozymes, the acyltransferases of the endoplasmic reticulum only produce PA with an 18-carbon fatty acid at the sn-2 position; 16:0, when present, is confined to the sn-1 position.[8] This PA gives rise to the phospholipids such as phosphatidylcholine (PC), phosphatidylethanolamine (PE) and phosphatidylinositol (PI), which are characteristic of the various extrachloroplast membranes. In addition, however, the diacylglycerol moiety of PC is returned to the

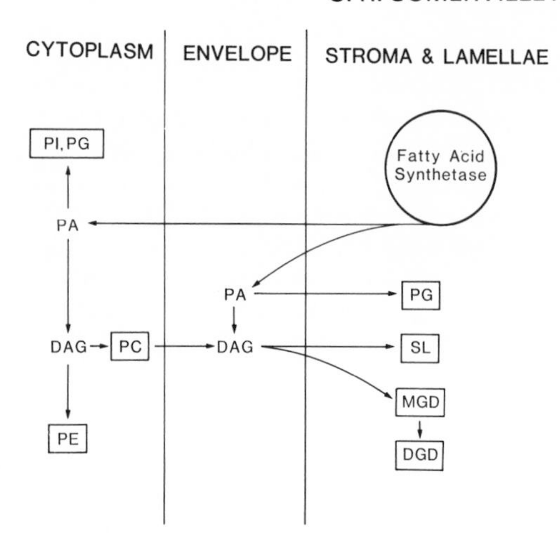

Fig. 1. An abbreviated diagram of the two-pathway scheme of glycerolipid biosynthesis in the 16:3 plant Arabidopsis. Fatty acids synthesized in the chloroplast are esterified to glycerol-3-phosphate by distinct acyltransferases in the plastid envelope or the endoplasmic reticulum to give rise to two separate pools of phosphatidic acid (PA). The PA in the chloroplast envelope is used for the synthesis of phosphatidylglycerol (PG) or converted to diacylglycerol (DAG). In the endoplasmic reticulum, PA is used directly to synthesize PG and phosphatidylinositol (PI) or via DAG to synthesize phosphatidylcholine (PC) and phosphatidylethanolamine (PE). In all higher plants a large proportion of the PC produced in the endoplasmic reticulum is used to provide a second supply of lipid molecules to the pool of DAG in the chloroplast envelope. This DAG is then used for production of sulfolipid (SL), monogalactosyl diacylglycerol (MGD) and digalactosyl diacylglycerol (DGD).

chloroplast envelope where it enters the DAG pool and contributes to the synthesis of thylakoid lipids (Fig. 1).[2,9]

In many species of higher plants PG is the only product of the prokaryotic pathway and the remaining chloroplast lipids are synthesized entirely by the eukaryotic pathway. In those species, such as Arabidopsis, in which both pathways do contribute to the synthesis of MGD, DGD and SL, the leaf lipids characteristically contain substantial amounts of hexadecatrienoic acid (16:3) which is found only in MGD and DGD molecules produced by the prokaryotic pathway. These plants have been termed 16:3 plants to distinguish them from the other angiosperms (18:3 plants) whose galactolipids contain predominantly linolenate.[10] The contribution of the eukaryotic pathway to MGD, DGD and SL synthesis is reduced in lower plants, and in many green algae the chloroplast is almost entirely autonomous with respect to membrane lipid synthesis.

A detailed quantitative analysis of the relative contributions of the two pathways was carried out in the 16:3 plant *Arabidopsis thaliana*.[11] These studies indicated that approximately 38% of newly synthesized fatty acids enter the prokaryotic pathway of lipid biosynthesis. Of the 62% which is exported as acyl-CoA species to enter the eukaryotic pathway, 56% (34% of the total) is ultimately reimported into the chloroplast. Thus, chloroplast lipids of *Arabidopsis* are about equally derived from the two pathways. The results of these studies are schematically presented in Figure 2.

FATTY ACID DESATURATION

The membrane lipids of higher plants contain primarily 16-carbon and 18-carbon acyl groups with 0 to 3 *cis* double bonds.[12] As a rule, the first unsaturation is at the n-9 position, the second is at the n-6 and the third is at the n-3 position (i.e., 3 carbons from the methyl end of the acyl group). The major exception to this is the presence of a *trans* unsaturation which occurs uniquely at the n-12 (i.e., Δ3) position of the 16-carbon acyl group on the *sn*-2 position of chloroplast PG. It is not known with any degree of certainty how many enzymes participate in the desaturation of plant lipids. The only plant desaturase that has been characterized in any detail is the soluble steroyl-ACP desaturase which inserts a double bond at the n-9 position in 18:0-ACP. This enzyme has been highly purified from developing safflower seeds and shown to require oxygen and reduced ferredoxin.[13] It is essentially inactive with palmitoyl-ACP (16:0-ACP) as a substrate. No other chloroplast desaturase has been assayed, although it is possible to detect the formation of linolenic acid (18:3) in isolated spinach chloroplasts.[14] From these

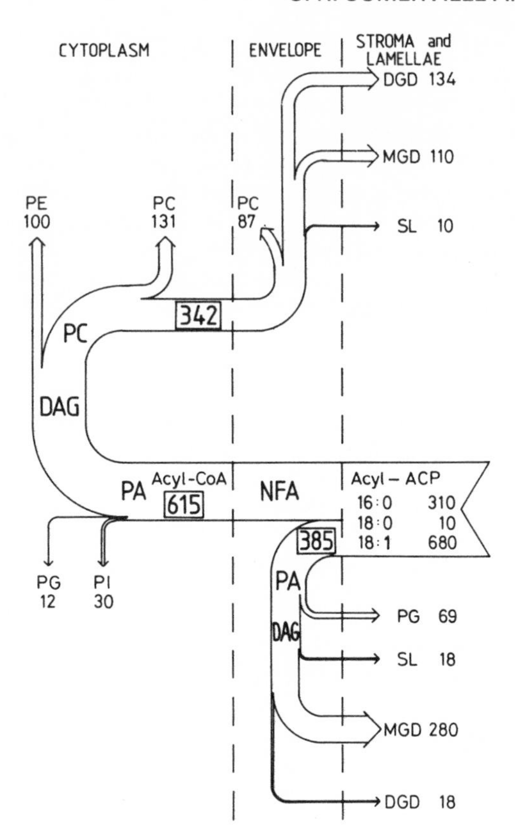

Fig. 2. A flow diagram showing the fluxes (mol/1000 mol) of fatty acids during lipid synthesis by _Arabidopsis_ leaves. The abbreviations for the various lipids are as in Figure 1. NFA indicates non-esterified fatty acids. Adapted from Reference 11.

studies it appeared that the substrate for the chloroplast desaturases was MGD.

By contrast with the lability of most desaturases, an NAD(P)H-dependent oleoyl phosphatidylcholine desaturase activity can be measured in the microsomal fraction from pea leaves[15] and developing safflower seeds.[16] A linoleate desaturase has also been demonstrated in microsomes from developing linseed cotyledons[17] and from soybean cotyledons.[18] However, it has not been possible to maintain activity of the microsomal enzymes in detergent solubilized preparations. In vertebrates the microsomal desaturase utilizes stearyl-CoA as substrate and requires cytochrome b_5 and cytochrome b_5 reductase as cofactors.[19] Thus, it seems possible that in order to purify the plant microsomal desaturases it will first be necessary to identify the cofactors and prepare them in amounts which will permit reconstitution of a complete system as has been done with the enzymes from vertebrates (reviewed in Reference 19). The recent purification of cytochrome b_5 from potato tubers[20] and the purification of cytochrome b_5 reductase[21,22] are important steps in this direction.

Most of the double bonds in membrane lipids are cis. By contrast, the 16-carbon at the sn-2 position of PG is found only as 16:0 or as trans-Δ3-16:1. Chloroplast PG is the only lipid in which the trans-Δ3-16:1 acyl group is found.[23] Because of these unusual characteristics, it has been suggested that trans-16:1-PG plays an important, even essential, role in photosynthesis.

MUTATIONS AFFECTING SEED LIPIDS

The biosynthetic pathways involved in the synthesis of seed storage lipids are similar or identical in most important respects to the pathways which are responsible for the synthesis of membrane lipids (reviewed in Reference 24). Thus, information gained from the analysis of mutants affecting the biosynthesis of seed lipid is relevant to membrane lipid biosynthesis and vice versa. This is likely to be particularly important in the context of isolating the genes which regulate the composition of seed lipid. Although, as noted below, different isozymes may be active in seeds and leaves, it seems possible that genes for leaf isozymes will be useful probes for genes of seed isozymes and vice versa.

Most of the genetic studies of plant lipid composition have been initiated by plant breeders and have, therefore, focused on alterations in seed lipid composition. Natural or induced variation has been observed in the desaturation of stearic, oleic and linoleic acid in safflower,[25] maize,[26,27] flax,[28] rapeseed,[29,30] soybean[31-34] and sunflower.[35] Many of these variants have substantial changes in fatty acid composition which have not been characterized in sufficient detail to be informative about the nature of the genetic alteration. An exception is the isolation of two chemically induced mutants of flax (Linum usitatissimum), each of which contained about 50% of levels of linolenic acid (18:3) in the wild type and had corresponding increases in 18:2 content.[28] When the two mutations were combined to make the double mutant, a line almost completely deficient in 18:3 in seed lipids was obtained. Analysis of the lipid composition of the double mutant[36] indicated that the large reduction in desaturation was not restricted to triacylglycerols but affected glycolipids and phospholipids to a similar extent. However there was apparently no effect on leaf lipids or on the viability of the seeds. The simplest interpretation of these results is that there are two genes encoding two desaturases in the seeds and that the products of these genes have been inactivated by the mutations. Since there was little or no effect on the composition of leaf fatty acid, it can be suggested that there are additional desaturase genes which are leaf specific. However, the enzymological analysis was not done.

A particularly interesting mutant (A6) of soybean was identified following chemical mutagenesis.[33] This line has a recessive nuclear mutation that caused an increase in stearic acid from about 4% of total seed lipid to about 28%. The increase in stearic acid content was accompanied by a corresponding decrease in oleic acid content, but there was no effect on palmitic, linoleic or linolenic acid content. Since oleic acid is the precursor of linoleic and linolenic acids, it is difficult to imagine that this mutant has a defect in the 18:0-desaturase. The biochemical analysis of this mutant should, therefore, prove very interesting.

Another soybean mutant (A5) exhibits a 2-fold increase in the amount of 18:1 in seed lipids. Analysis of the species composition of PC and PE indicated that all positions are equally affected by the mutation.[34] Assay of desaturase activity in microsomes of this mutant revealed a 35% decrease in the amount of the oleoyl-PC desaturase activity in the mutant. Presumably the mutation is either leaky or there are several genes for the 18:1-desaturase in soybean.

It is apparent that in spite of the substantial amount of effort which has been invested in screening diverse genotypes of various species, it is only recently that the variants have been characterized in any detail. None of the mutants appears to have been exploited as an experimental system to learn something about the underlying mechanisms responsible for regulation of seed lipid composition. Hopefully this situation is changing.

The only significant change in seed lipid composition not based primarily on changes in the level of fatty acid unsaturation is the development of rapeseed cultivars containing low levels of erucic (22:1) and eicosenoic (20:1) acids.[37]

MUTANTS WITH ALTERED LEAF LIPIDS

The isolation of a series of mutants with defects in leaf lipid metabolism was originally motivated by the concept that, in order to study the functional significance of lipid unsaturation, it would be very useful to have a collection of otherwise isogenic mutant lines which differ only with respect to the activity of specific desaturases. We chose the small crucifer *Arabidopsis thaliana* (L.) as the experimental organism because it has a number of traits which render it particularly well-suited for physiological genetics.[38]

Since there was no obvious way to identify mutants with altered lipid composition on the basis of gross phenotype, we screened for mutants by direct assay of fatty acid composition of leaf tissue. To facilitate this approach we devised a simple and rapid procedure for preparing fatty acid methyl esters from small samples of fresh leaf tissue.[39] Using this method it is possible to obtain quantitative information on the fatty acyl composition of the lipids from as little as 5 mg of leaf tissue within several hours. We then employed this method to measure the fatty acyl composition of total lipids from single leaves from randomly chosen plants in a population of *Arabidopsis* that had been mutagenized with ethylmethane sulfonate. From among the first 2000 M2 plants examined in this way we identified 7 mutants with major changes in fatty acyl composition,[40] and in subsequent searches we have identified several additional mutant lines. The fatty acid

Table 1. Fatty acid composition of leaf lipids from some mutants of *Arabidopsis*.

Mutant Line	Fatty Acid Composition								
	16:0	16:1	16:1t	16:2	16:3	18:0	18:1	18:2	18:3
Wild type	16	tr	2	tr	12	1	4	18	47
JB60	18	tr	0	tr	12	1	3	19	47
JB1	17	3	3	6	2	1	9	39	19
LK3	15	8	2	-	-	1	21	18	34
JB67	24	-	2	-	-	1	2	14	56
JB25	10	tr	2	tr	tr	1	8	23	54

compositions of leaves from 5 mutants are shown in Table 1. In all of these mutants the alteration in fatty acid composition has been found to be due to single, recessive nuclear mutations. The sites of the enzymatic lesions in four of these mutants (designated *fadA*, *fadB*, *fadC*, *fadD*) have been tentatively determined as illustrated in Figure 3.

The first mutant to be characterized in detail was completely lacking the acyl group *trans*-hexadecenoic acid due to a mutation at a locus designated *fadA*.[40] This acyl group normally occurs only on the *sn*-2 position of phosphatidylglycerol (PG) in chloroplast membranes.[23] There were no other changes in the fatty acid composition of this mutant except for a corresponding increase in the amount of 16:0. Thus, it was inferred that the *fadA* locus encodes a desaturase which acts specifically on 16:0 at the *sn*-2 position of PG.

The distinguishing characteristic of a mutant with a lesion at the *fadD* locus is a substantial decrease in the amount of both 16:3 and 18:3 fatty acids in extracts of whole leaves and a corresponding increase in the amount of 16:2 and 18:2, respectively.[41] This is consistent with the conclusion that the *fadD* locus controls the activity of a

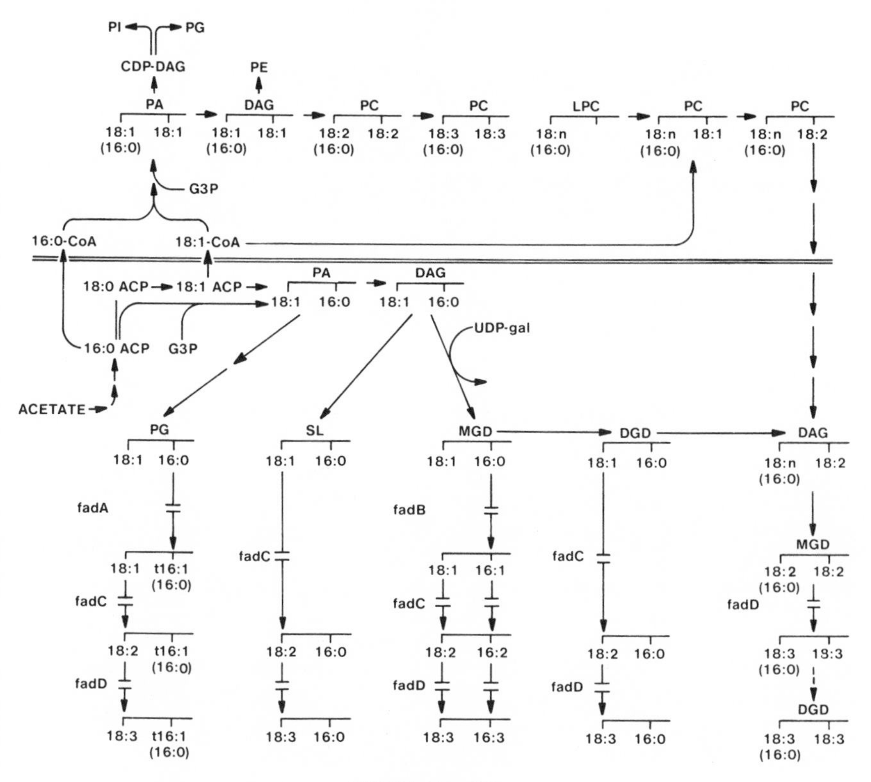

Fig. 3. An abbreviated scheme for glycerolipid synthesis in *Arabidopsis* leaves showing the proposed locations of the enzymatic deficiencies in four mutants with altered fatty acid composition. The lesions are indicated by a break in the diagram adjacent to the gene symbol (i.e., *fadA*, *fadB*, *fadC* and *fadD*).

desaturase which converts both 16- and 18-carbon dienoic to trienoic acyl groups. Analysis of the molecular species of MGD and DGD in the mutant and the wild-type indicated that the mutation affects the desaturation of fatty acids at both the *sn*-1 and *sn*-2 position of MGD and DGD.[42]

The effect of the fadD mutation was fully expressed only when plants were grown at temperatures above about 28°C. The simplest hypothesis to explain this is that the mutation renders the desaturase temperature sensitive. However, since there are other conceivable explanations for this effect we are searching for additional mutant alleles of the fadD locus in order to determine if mutations at this locus always render the amount of trienoic acid responsive to temperature.

One of the unexpected and intriguing properties of the mutant is that the amount of 18:3 is reduced to a similar extent (about 45% reduction at temperatures above about 28°C) in all the major lipids (i.e., MGD, DGD, SL, PG, PC, PE, PI). Since 16:3 is thought to be synthesized from 16:2 only in the chloroplast, the desaturase must be located within the chloroplast. However, there is no PE and only very low levels of PI in the chloroplast.[43] Thus, it is apparent that either the same desaturase is located in both compartments, or that the 18:3 on PE and the other cytoplasmic lipids is synthesized in the chloroplast and transported to the cytoplasm. This could involve the transfer of 18:3 as a CoA ester (Fig. 2) or perhaps by reversal of the process which transfers chloroplast fatty acids to the eukaryotic pathway. In view of the fact that transfer of acyl groups between compartments is already known to occur in plants,[2,44,45] we favor the latter possibility. However, the recent observation that the same gene encodes both the cytoplasmic and mitochondrial histidine tRNA synthetase of yeast[45] indicates that the former explanation is also a possibility.

In addition to the fadA and fadD mutants we have isolated mutants (designated fadC in Fig. 1) which appear to be deficient in the chloroplast desaturase which converts 16:1 and 18:1 to 16:2 and 18:2, respectively. We have also isolated mutants (fadB) which accumulate high levels of 16:0 but which contain no 16:1, 16:2 or 16:3 fatty acids. A third class of as yet uncharacterized mutants are deficient in 16:3 but do not have increased amounts of the more saturated 16-carbon fatty acids, and have slightly increased levels of all 18-carbon fatty acids. Thus, we infer that this class of mutants is not deficient in a desaturase per se but affects the partitioning of fatty acids between the prokaryotic and eukaryotic pathways. Preliminary results indicate that these mutants are deficient in the acyltransferase which esterifies 18:1 to position sn-1 of glycerol-3-phosphate in the chloroplast (unpublished results).

PHYSIOLOGICAL ROLE OF LIPID UNSATURATION

Chloroplast membranes have a characteristic and unusual fatty acid composition. Typically linolenic acid (18:3) or a combination of linolenic and hexadecatrienoic acid (16:3) account for approximately two-thirds of all the fatty acids of thylakoid membranes and over 90% of the fatty acids of monogalactosyldiacylglycerol (MGD), the most abundant chloroplast lipid.[47] The atypical fatty acid trans-Δ3-hexadecenoate (trans-16:1) is present as a component of phosphatidylglycerol (PG), the major thylakoid phospholipid. The fact that these and other characteristics of chloroplast lipids are common to most or all higher plant species suggests that the lipid fatty acid composition is important for maintaining photosynthetic function. Furthermore, the thylakoid membrane is the site of light absorbtion and oxygen production. The free radicals which are by-products of these reactions will stimulate oxidation of the polyunsaturated fatty acids. Since this might be expected to mediate against a high degree of unsaturation it may be inferred that there is a strong selective advantage to having such high levels of trienoic fatty acids in thylakoid membranes.

Many different approaches have been used in attempts to elucidate the significance of membrane fatty acid composition to photosynthetic function. These include the correlation of events during chloroplast development,[48,49] physical studies of model membrane systems,[50-52] reconstitution of photosynthetic components with lipid mixtures,[53] alteration of lipids in situ by heat stress,[54] lipase treatment,[55] chemical inhibitors,[56] or hydrogenation of unsaturated fatty acids.[52,57,58] To date however, these approaches have not been successful in establishing any unequivocal relationships between membrane form and function.

The availability of a series of biochemically characterized mutants of Arabidopsis with specific alterations in leaf membrane fatty acid composition offers a novel approach to the general problem of the role of lipid polymorphism.[40] Here we review the ways in which several of these mutants have been used to test particular hypotheses about the functional significance of thylakoid fatty acid composition.

A Mutant Lacking trans-16:1

trans-16:1 Is an atypical fatty acid because of the position of the double bond proximal to the carboxyl group and because of the *trans* configuration. The almost ubiquitous occurrence of *trans*-16:1 in photosynthetic membranes[23] and the fact that its synthesis has been strongly correlated with the development of photosynthetic competence[49,59] led to the assumption that this fatty acid had an important or essential function in photosynthesis. Early attempts to define a role in granal stacking could not be sustained.[59] Alternatively, experiments in which *trans*-16:1 was removed by treatment of isolated thylakoids with phospholipase A2 resulted in changes in the efficiency of light capture and the kinetics of fluorescence induction.[60] More recently, analysis of chlorophyll-protein complexes separated by SDS gel electrophoresis have shown that the band containing the oligomeric form of the light harvesting chlorophyll a/b protein complex (LHCP) is enriched in PG containing *trans*-16:1.[61] Furthermore, reconstitution of the LHCP oligomer from the monomeric form in liposomes was stimulated by the presence of *trans*-16:1.[62] These and other observations (reviewed in Ref. 23) have been taken to indicate that *trans*-16:1 is a structural component of oligomeric LHCP and that this constitutes its vital role in photosynthesis.

The fact that mutants lacking *trans*-16:1 were isolated as healthy vigorous plants[40] demonstrates that this fatty acid has no critical function in photosynthesis. In addition, we have investigated the mutant and wild-type more closely in order to specifically assess suggestions about the importance of *trans*-16:1 in thylakoid membranes. The results of these investigations were as follows:[40,63]

[1] There was no distinguishable difference in chloroplast ultrastructure between the mutant and wild-type;

[2] Infrared gas analysis of whole-plant photosynthesis and electron transport measurements with isolated thylakoids demonstrated no alteration of photosynthetic characteristics;

[3] There was no significant change in fluorescence induction kinetics or 77K fluorescence spectra of the mutant as compared to the wild-type;

[4] High temperature enhancement of the fluorescence yield from leaves of mutant and wild-type plants showed the same temperature response, indicating that the chlorophyll-protein complexes of the mutant were not more susceptible to thermal denaturation than were those of the wild-type.

[5] Despite the absence of any functional difference by these various criteria, we did find that the oligomeric form of LHCP was slightly more sensitive to dissociation during SDS extraction and gel electrophoresis as the NaCl concentration of the buffers was increased.[61]

This latter observation is consistent with trans-16:1 being involved in stabilizing the LHCP oligomer, but suggests that the effects must be very subtle. This conclusion is supported by the recent discovery that a number of orchid species also lack trans-16:1.[64] It is possible that the role of trans-16:1 is restricted to unusual environmental conditions or to a specific stage of development. For example, it may assist the insertion of LHCP into the developing thylakoids during chloroplast biogenesis. Whatever the precise role, it seems apparent that this fatty acid is an element of the fine tuning mechanisms which have evolved to optimize the efficiency of photosynthetic electron transport. Further study of this Arabidopsis mutant would appear to be the most realistic route for any further investigations of the role of trans-16:1 in chloroplast function. Thus, for instance, attempts to determine cause and effect relationships from correlations between changes in trans-16:1 content and cold-induced functional reorganization of chloroplast membranes[65] seem relatively equivocal.

A Mutant Deficient in Trienoic Fatty Acids

Speculations about the presumed importance of a highly unsaturated bilayer in thylakoid membranes has centered on three proposals. First, it has been suggested that the role of polyunsaturation is to maintain membrane fluidity. It is well established that the high degree of unsaturation provides a very fluid matrix even when the ordering effect of intrinsic membrane proteins is taken into account.[66] However, theoretical considerations and analysis of model systems indicate that the major increase in fluidity occurs with the insertion of the first *cis* double bond in the hydrocarbon chain. Insertion of additional double bonds has little or no effect on fluidity.[67]

Alternate hypotheses are based on detailed considerations of the properties of MGD, the major thylakoid lipid which also contains the highest proportions of trienoic fatty acids. When dispersed in water at room temperature, MGD isolated from higher plant leaves spontaneously forms inverted tubular structures, the hexII phase.[51,54] These structures do not occur *in vivo* at normal temperatures but their formation at higher temperatures has been implicated in the disruption of thylakoid membranes. Such disruption is thought to be related to the enhanced fluorescence yield from PSII which is observed in leaves or thylakoid preparations heated to 35-45°C[58] even though electron microscopic detection of hexII lipids requires rather higher temperatures.[54] The tendency of MGD to form hexII structures has been suggested to aid in the packing of large proteins into the thylakoid membrane. Conversely, the admixture of MGD with proteins and other components may prevent the conversion of MGD to the hexII form at normal physiological temperatures.[53]

In experiments in which MGD was progressively saturated by catalytic hydrogenation, it was found that a reduction in the average number of double bonds per molecule from 5.44 in the untreated leaf to 5.10 was sufficient to induce the formation of regions of lamellar structure while further hydrogenation increased the proportion of lamellar to hexII lipid.[53] Thus, one rationale for the high trienoic fatty acid content is to confer upon MGD the ability to stabilize protein complexes in the thylakoid membrane, an ability which is associated with the tendency to form hexII structures. Indirect support for this view has come from studies of *Nerium oleander* plants adapted to different temperature regimens. In these studies the temperature at which thermal enhancement of fluorescence yield was observed could be correlated with the extent of unsaturation of the thylakoid lipids.[68]

After formation of the lamellar phase further hydrogenation (to an average of 4.3 double bonds per molecule or less) resulted in a change of the lamellar lipid from the liquid crystalline to gel phase at room temperature.[53] Several lines of evidence indicate that such a phase transition, if it occurred *in vivo*, would seriously disrupt photosynthetic reactions occurring in the thylakoids.[52,53] To date however, experiments on the catalytic hydrogenation of chloroplast membranes *in vitro* and *in vivo* have failed to show any extensive effects of moderate decreases in the extent of lipid unsaturation.[52,53,56-58]

The trienoic fatty acid content of thylakoids has also been decreased by growing barley seedlings in sublethal concentrations of the herbicide SAN9785.[56] In these studies the ratio of PSII to PSI and the ratio of granal to stromal thylakoids were both increased but these effects may not be related to the change in fatty acid composition since the herbicide is a direct inhibitor of photosynthetic electron transport.[69]

We have investigated the photosynthetic characteristics of an *Arabidopsis* mutant which has large reductions in both 18:3 and 16:3 fatty acids and has concomitant increases in 18:2 and 16:2.[41] These changes have been attributed to a deficiency of an n-3 desaturase which normally desaturates both 18:2 and 16:2. Even though the average number of double bonds per MGD molecule is reduced from 5.75 in the wild-type to 4.67 in the mutant,[41] plants from the mutant line do not show any obvious phenotype, and grow at rates comparable to the wild-type. Apart from demonstrating in a general sense that high concentrations of trienoic acids are not required for normal plant development and growth, our studies on the mutant contradict some of the conclusions reached by the use of less specific approaches.

Examination of the behavior of the exogenous fluorescence probe diphenyl hexatriene showed no more than a very minor increase in thylakoid membrane fluidity between mutant and wild type.[70] Similarly, chlorophyll fluorescence characteristics were unchanged in the mutant. In particular, the temperature at which the thermal enhancement of fluorescence yield occurs was indistinguishable from wild-type.[70] Since the change in trienoic fatty acid composition in the mutant was greater than those which accompanied high temperature adaptation of *N. oleander* plants,[68] we conclude that the fatty acid composition of MGD is not the critical factor determining the temperature at which enhanced chlorophyll fluorescence occurs. We do not yet know whether the altered fatty acid composition has an effect on the formation of hexII lipid structures in the thylakoid membranes.

The most pronounced change in the mutant was a 45% reduction in the cross sectional area of chloroplasts. This was associated with an increase in chloroplast number, small changes in chloroplast morphology, and a net 15% decrease in the amount of chlorophyll per gram fresh weight. These changes appeared to be related to the alteration in fatty acid composition on the basis of several criteria, but they had little or no effect on a range of photosynthetic parameters (expressed on a chlorophyll basis) including net CO_2 exchange of whole plants and the PSI, PSII and whole chain electron transport rates of thylakoid preparations.[68]

Taken as a whole, these observations imply that the high trienoic acid content of thylakoid membranes may be related to chloroplast morphogenesis rather than directly to photosynthesis. However, in the absence of supporting data from studies with SAN9785 herbicide[56] or a mechanistic hypothesis to relate chloroplast morphology to fatty acid composition, it would be instructive to have further information from independently derived mutants.

CONCLUSIONS

It appears that plant lipid metabolism is amenable to experimental manipulation by the isolation of a wide spectrum of mutants. The mutants described here provide useful tools for studies of the regulation and functional significance of lipid unsaturation. The anticipated development of more advanced methods of genetic analysis in *Arabidopsis*[38] may also facilitate the isolation of the genes affected by the mutations. Indeed, because the desaturase enzymes have been intractable, it seems possible that the isolation of the genes for the desaturases may be useful in understanding these enzymes. The mutants may be useful in order to clone the genes without first purifying the desaturases. For instance, some of them may have missing or altered polypeptides which can be used to identify the wild-type polypeptide. Alternatively, it may be possible to walk into the genes by chromosome walking or in certain cases to isolate the genes by shotgun complementation.

A major conclusion from the mutant analysis which has been completed to date is that plants can withstand relatively large changes in membrane and storage lipid composition without serious effects. However, it seems likely that there are many changes in lipid composition which would be incompatible with the survival of the plant. In particular, changes in the amount of the various lipid headgroups could be severely deleterious. An alternative approach to the analysis of questions concerning the role of head groups may be to effect quantitative changes in the amount of specific lipids by introducing cloned genes into transgenic plants. On the one hand it should be possible to increase the amount of a lipid by increasing the amount of activity of the rate-limiting steps. For example, all the desaturase mutants discussed in this review show gene dosage effects, as does the gene that controls fatty acid elongation in rapeseed. This implies that the extent of desaturation may be limited by the extent of expression of the desaturase genes. Thus, over-expression of the cloned genes in transgenic plants may lead to more highly unsaturated oils. Conversely, it may be possible by techniques such as antisense mRNA to reduce the amount of key enzymes. If the lipids are essential for some process, it should be possible to examine the roles of lipids by affecting such quantitative changes without killing the plants.

In contrast to many alternative approaches, mutant analysis offers the potential to provide clear and unequivocal information about how lipid composition affects plant function. Once a single mutation has been established in an otherwise uniform wild-type genetic background, then it follows that all the differences between the mutant and the wild-type must be related directly to the mutation. The mutants described here are particularly useful in this respect since they have no readily apparent phenotype. Thus, in contrast to the situation where one cannot distinguish between primary and secondary effects of a mutation, we have simply been unable to identify any major effects on membrane function. We have, therefore, been able to convincingly exclude a number of hypotheses which have been previously proposed.[40,61,69]

The effects of the mutations on chloroplast biogenesis are suggestive but need to be substantiated by the isolation and characterization of additional independent mutants with similar changes in fatty acid composition. We anticipate that these studies, in conjunction with the characterization of the other classes of mutants which we have not yet examined in detail, will facilitate the formulation of new hypotheses concerning the significance of fatty acid unsaturation to membrane structure, function and biogenesis.

ACKNOWLEDGMENTS

This work was supported in part by grants (#DE-AC02-76ER01338) from the U.S. Department of Energy and the McKnight Foundation.

REFERENCES

1. LYONS, J.M., J.K. RAISON, P.L. STEPONKUS. 1979. The plant membrane in response to low temperature. In Low Temperature Stress in Crop Plants: The Role of the Membrane. (J.M. Lyons, D. Graham, J.K. Raison, eds.), Academic Press, New York, pp. 1-24.
2. ROUGHAN, P.G., C.R. SLACK. 1982. Cellular organization of glycerolipid metabolism. Annu. Rev. Plant Physiol. 33: 97-123.
3. FRENTZEN, M., E. HEINZ, T.A. McKEON, P.K. STUMPF. 1983. Specificities and selectivities of glycerol-3-phosphate acyltransferase from pea and spinach chloroplasts. Eur. J. Biochem. 129: 629-636.
4. BLOCK, M.A., A.J. DORNE, J. JOYARD, R. DOUCE. 1983. The phosphatidic acid phosphatase of the chloroplast envelope is located on the inner envelope membrane. FEBS Lett. 164: 111-115.
5. ANDREWS, J., J.B. OHLROGGE, K. KEEGSTRA. 1985. Final step of phosphatidic acid synthesis in pea chloroplasts occurs in the inner envelope membrane. Plant Physiol. 78: 459-466.
6. HEEMSKERK, J.W.M., G. BOGEMANN, J.F.G.M. WINTERMANS. 1985. Spinach chloroplasts: localization of enzymes involved in galactolipid metabolism. Biochim. Biophys. Acta 835: 212-220.
7. COVES, J., M.A. BLOCK. J. JOYARD, R. DOUCE. 1986. Solubilization and partial purification of UDP-galactose diacylglycerol galactosyl transferase activity from spinach chloroplast envelope. FEBS Lett. 208: 401-407.

8. FRENTZEN, M., W. HARES, A. SCHIBURR. 1984. Properties of the microsomal glycerol-3-P and monoacylglycerol-3-P acyltransferases from leaves. In Structure, Function and Metabolism of Plant Lipids. (P.A. Siegenthaler, W. Eichenburger, eds.), Elsevier, Amsterdam, pp. 105-110.
9. HEINZ, E., P.G. ROUGHAN. 1983. Similarities and differences in lipid metabolism of chloroplasts isolated from 18:3 and 16:3 plants. Plant Physiol. 72: 273-279.
10. JAMIESON, G.R., E.H. REID. 1971. The occurrence of hexadeca-7,10,13-trienoic acid in the leaves of angiosperms. Phytochemistry 10: 1837-1843.
11. BROWSE, J.A., N. WARWICK, C.R. SOMERVILLE, C.R. SLACK. Fluxes through the prokaryotic and eukaryotic pathway of lipid synthesis in the 16:3 plant Arabidopsis thaliana. Biochem. J. 235: 25-31.
12. FRENTZEN, M. 1986. Biosynthesis and desaturation of the different diacylglycerol moieties in higher plants. J. Plant Physiol. 124: 193-209.
13. McKEON, T.A., P.K. STUMPF. 1982. Purification and characterization of the stearoyl-acyl carrier protein desaturase and the acyl-acyl carrier protein thioesterase from maturing seeds of safflower. J. Biol. Chem. 257: 12141-12147.
14. ROUGHAN, P.G., J.B. MUDD, T.T. McMANUS, C.R. SLACK. 1979. Linoleate and α-linolenate synthesis by isolated chloroplasts. Biochem. J. 184: 571-574.
15. MURPHY, D.E., I.E. WOODROW, E. LATZKO, K.D. MUKHERJEE. 1983. Solubilization of oleoyl-CoA thioesterase, oleoyl-CoA:PC acyltransferase and oleoyl phosphatidylcholine desaturase. FEBS Lett. 162: 442-446.
16. GENNITY, J.M., P.K. STUMPF. 1985. Studies of the Δ12 desaturase of Carthamus tinctorius L. Arch. Biochem. Biophys. 239: 444-454.
17. BROWSE, J.A., C.R. SLACK. 1981. Catalase stimulates linoleate desaturase activity in microsomes from developing linseed cotyledons. FEBS Lett. 131: 111-114.
18. STYMNE, S., L.A. APPELQVIST. 1980. The biosynthesis of linoleate and α-linolenate in homogenates from developing soybean cotyledons. Plant Sci. Lett. 17: 287-294.
19. HOLLOWAY, P.W. 1983. Fatty acid desaturation. The Enzymes 16: 63-83.

20. BONNEROT, C., A.M. GALLE, A. JOLLIOT, J.C. KADER. 1985. Purification and properties of plant cytochrome b_5. Biochem. J. 226: 331-334.
21. GALLE, A.M., C. BONNEROT, A. JOLLIOT, J.C. KADER. 1984. Purification of a NADH-ferricyanide reductase from plant microsomal membranes with zwitterionic detergent. Biochem. Biophys. Res. Commun. 122: 1201-1205.
22. GALLE, A.M., J.C. KADER. 1986. High performance liquid chromatography of plant membrane proteins. NADH-cytochrome b_5 reductase as a model. J. Chromatogr. 366: 422-426.
23. DUBACQ, J.P., A. TREMOLIERES. 1983. Occurrence and function of phosphatidylglycerol containing _trans_-Δ3-hexadecenoic acid in photosynthetic lamellae. Physiol. Veg. 21: 293-312.
24. SLACK, C.R., J.A. BROWSE. 1984. Synthesis of storage lipids in developing seeds. _In_ Seed Physiology. (D.M. Murray, ed.), Vol. _1_, Academic Press, New York, pp. 209-243.
25. KNOWLES, P.F., A.B. HILL. 1964. Inheritance of fatty acid content in the seed oil of a safflower introduction from Iran. Crop Sci. 4: 406-409.
26. PONELEIT, C.G., D.E. ALEXANDER. 1965. Inheritance of linolenic acid and oleic acid in maize. Science 147: 1585-1586.
27. WIDSTROM, N.W., M.D. JELLUM. 1984. Chromosomal location of genes controlling oleic and linoleic acid composition in the germ oil of two maize inbreds. Crop Sci. 24: 1113-1115.
28. GREEN, A.G., D.R. MARSHAL. 1984. Isolation of induced mutants in linseed (_Linum usitatissimum_) having reduced linolenic acid content. Euphytica 33: 321-328.
29. ROBBELEN, G., A. NITSCH. 1975. Genetical and physiological investigations on mutants for polyenoic fatty acids in rapeseed _Brassica napus_ L. Z. Pflanzenzuecht. 75: 93-105.
30. DIEPENBROCK, W., R.F. WILSON. 1987. Genetic regulation of linolenic acid concentration in rapeseed. Crop Sci. 27: 75-77.
31. WILCOX, J.R., J.F. CAVINS, N.C. NIELSEN. 1984. Genetic alteration of soybean oil composition by a chemical mutagen. J. Am. Oil Chem. Soc. 61: 97-100.

32. WILCOX, J.R., J.F. CAVINS. 1985. Inheritance of low linolenic acid content of the seed coat of a mutant in Glycine max. Theor. Appl. Genet. 71: 74-78.
33. GRAEF, G.L., L.A. MILLER, W.R. FEHR, E.G. HAMMOND. 1985. Fatty acid development in a soybean mutant with high stearic acid. J. Am. Oil Chem. Soc. 62: 773-775.
34. MARTIN, B.A., R.W. RINNE. 1986. A comparison of oleic acid metabolism in the soybean (Glycine max [L] Merr.) genotypes Williams and A5, a mutant with decreased linoleic acid in the seed. Plant Physiol. 81: 41-44.
35. PURDY, R.H. 1986. High oleic sunflower: physical and chemical characteristics. J. Am. Oil Chem. Soc. 63: 1062-1065.
36. TONNET, M.L., A.G. GREEN. 1987. Characterization of the seed and leaf lipids of high and low linolenic acid flax genotypes. Arch. Biochem. Biophys. 252: 646-654.
37. DOWNEY, R.K., D.I. McGREGOR. 1975. Breeding for modified fatty acid composition. Curr. Adv. Plant Sci. 12: 151-167.
38. ESTELLE, M.A., C.R. SOMERVILLE. 1986. The mutants of Arabidopsis. Trends Genet. 2: 89-93.
39. BROWSE, J., P. McCOURT, C.R. SOMERVILLE. 1985. Overall fatty acid composition of leaf lipids determined after combined digestion and fatty acid methyl-ester formation from fresh tissue. Anal. Biochem. 152: 141-146.
40. BROWSE, J., P. McCOURT, C.R. SOMERVILLE. 1985. A mutant of Arabidopsis lacking a chloroplast specific lipid. Science 227: 763-765.
41. BROWSE, J., P. McCOURT, C.R. SOMERVILLE. 1986. A mutant of Arabidopsis deficient in C18:3 and C16:3 leaf lipids. Plant Physiol. 81: 859-864.
42. NORMAN, H.A., J.B. ST. JOHN. 1986. Metabolism of unsaturated monogalactosyl diacylglycerol molecular species in Arabidopsis thaliana reveals different sites and substrates for linolenic acid synthesis. Plant Physiol. 81: 731-736.
43. DOUCE, R., J. JOYARD. 1980. Chloroplast envelope lipids: detection and biosynthesis. Methods Enzymol. 69: 290-301.
44. YAMADA, M., J.I. OHNISHI. 1982. Glycerolipid synthesis in Avena leaves during greening of

etiolated seedlings III. Synthesis of α-linolenoyl monogalactosyl diacylglycerol from liposomal linoleoyl phosphatidylcholine by Avena plastids in the presence of phosphatidylcholine exchange protein. Plant Cell Physiol. 23: 767-773.

45. SLACK, C.R., P.G. ROUGHAN, N. BALASINGHAM. 1977. Labelling studies in vivo on the metabolism of the acyl and glycerol moieties of the glycerolipids in the developing maize leaf. Biochem. J. 162: 289-296.

46. NATSOULIS, G., F. HILGER, G.R. FINK. 1986. The HTS1 gene encodes both the cytoplasmic and mitochondrial histidine tRNA synthetase of S. cerevisiae. Cell 46: 235-243.

47. GOUNARIS, K., J. BARBER. 1983. Monogalactosyldiacylglycerol: the most abundant polar lipid in nature. Trends Biochem. Sci. 8: 378-381.

48. VAN WALRAVEN, H.S., E. KOPPENAAL, H.J.P. MARVIN, M.J.M. HAGENDOORN, R. KRAAYENHOF. 1984. Lipid specificity for the reconstitution of well coupled ATPase proteoliposomes and a new method for lipid isolation from photosynthetic membranes. Eur. J. Biochem. 144: 563-566.

49. LEECH, R.M., M.G. RUMSBY, W.W. THOMSON. 1973. Plastid differentiation, acyl lipid, and fatty acid changes in developing green maize leaves. Plant Physiol. 52: 240-245.

50. GALEY, J., B. FRANCKE, J. BAHL. 1980. Ultrastructure and lipid composition of etioplasts in developing dark-grown wheat leaves. Planta 149: 433-439.

51. SEN, A., W.P. WILLIAMS, P.J. QUINN. 1981. The structure and thermotropic properties of pure 1,2-diacylgalactosylglycerols in aqueous systems. Biochim. Biophys. Acta 663: 380-389.

52. QUINN, P.J., W.P. WILLIAMS. 1983. The structural role of lipids in photosynthetic membranes. Biochim. Biophys. Acta 737: 223-266.

53. GOUNARIS, K., D.D. MANNOCK, A. SEN, A.P.R. BRAIN, W.P. WILLIAMS, P.J. QUINN. 1983. Polyunsaturated fatty acid residues of galactolipids are involved in the control of bilayer/non-bilayer lipid transitions in higher plant chloroplasts. Biochim. Biophys. Acta 732: 229-242.

54. GOUNARIS, K., A.P.R. BRAIN, P.J. QUINN, W.P. WILLIAMS. 1984. Structural reorganisation of chloroplast

thylakoid membranes in response to heat stress. Biochim. Biophys. Acta 766: 198-208.

55. RAWYLER, A., P.A. SIEGENTHALER. 1981. Transmembrane distribution of phospholipids and their involvement in electron transport as revealed by phospholipase A2 treatment of spinach thylakoids. Biochim. Biophys. Acta 635: 348-368.
56. LEECH, R.M., C.A. WALTON, N.R. BAKER. 1985. Some effects of 4-chloro-5-dimethylamino-2-phenyl-3(2H)-pyridazinone (SAN9785) on the development of thylakoid membranes in Hordeum vulgare L. Planta 165: 277-283.
57. VIGH, L., F. JOO, M. DROPPA, L.I. HORVATH, G. HORVATH. 1985. Modulation of chloroplast membrane lipids by homogeneous catalytic hydrogenation. Eur. J. Biochem. 147: 477-481.
58. THOMAS, P.G., P.J. DOMINY, L. VIGH, A.R. MANSOURIAN, P.J. QUINN, W.P. WILLIAMS. 1986. Increased thermal stability of pigment-protein complexes of pea thylakoids following catalytic hydrogenation of membrane lipids. Biochim. Biophys. Acta 849: 131-140.
59. LEMOINE, Y., J.P. DUBACQ, G. ZABULON. 1982. Changes in light harvesting capacities and trans-Δ3-hexadecenoic acid content in dark- and light-grown Picea abies. Physiol. Veg. 20: 487-503.
60. DUVAL, J.C., A. TREMOLIERES, J.P. DUBACQ. 1979. The possible role of trans-hexadecenoic acid and phosphatidylglycerol in the light reactions of photosynthesis. FEBS Lett. 106: 414-418.
61. TREMOLEIRES, A., J.P. DUBACQ, F. AMBARD-BRETTEVILLE, R. REMY. 1981. Lipid composition of chlorophyll protein complexes. FEBS Lett. 130: 27-31.
62. REMY, R., A. TREMOLIERES, F. AMBARD-BRETTEVILLE. 1984. Formation of oligomeric light harvesting chlorophyll a/b protein by interaction between its monomeric form and liposomes. Photobiochem. Photobiophys. 7: 267-276.
63. McCOURT, P., J. BROWSE, J. WATSON, C.J. ARNTZEN, C.R. SOMERVILLE. 1985. Analysis of photosynthetic antenna function in a mutant of Arabidopsis thaliana (L.) lacking trans-hexadecenoic acid. Plant Physiol. 78: 853-858.
64. ROUGHAN, P.G. 1986. A simplified isolation of phosphatidylglycerol. Plant Sci. 43: 57-62.

65. HUNER, N.P., M. KROL, J.P. WILLIAMS, E. MAISSAN, P.S. LOW, D. ROBERTS, J.E. THOMPSON. 1987. Low temperature development induces a specific decrease in trans-Δ3-hexadecenoic acid content. Plant Physiol. 84: 12-18.

66. BARBER, J. 1983. Photosynthetic electron transport in relation to thylakoid membrane composition and organization. Plant Cell Environ. 6: 311-322.

67. SMALL, D.M. 1986. In The Physical Chemistry of Lipids: From Alkanes to Phospholipids. Plenum Press, New York, 665 pp.

68. RAISON, J.K., J.K.M. ROBERTS, J.A. BERRY. 1982. Correlation between the thermal stability of chloroplast (thylakoid) membranes and the composition and fluidity of their polar lipids upon acclimation of the higher plant Nerium oleander to growth temperature. Biochim. Biophys. Acta 688: 218-228.

69. HILTON, J.L., A.L. SCHAREN, J.B. ST. JOHN, D.E. MORELAND, K.H. NORRIS. 1969. Modes of action of pyridazinone herbicides. Weed Sci. 17: 541-547.

70. McCOURT, P., L. KUNST, J. BROWSE, C.R. SOMERVILLE. 1987. The effects of reduced amounts of lipid unsaturation on chloroplast ultrastructure and photosynthesis in a mutant of Arabidopsis. Plant Physiol. 84: 353-360.

Chapter Three

STUDIES ON THE 5-ENOLPYRUVYLSHIKIMATE-3-PHOSPHATE SYNTHASE GENES OF HIGHER PLANTS AND ENGINEERING OF GLYPHOSATE RESISTANCE

CHARLES S. GASSER, DILIP M. SHAH,
GUY DELLA-CIOPPA, STEPHEN M. PADGETTE,
GANESH M. KISHORE, HARRY J. KLEE,
STEPHEN G. ROGERS, ROBERT B. HORSCH
AND ROBERT T. FRALEY

Plant Molecular Biology
Monsanto Company
700 Chesterfield Village Parkway
St. Louis, Missouri 63198

INTRODUCTION

5-Enolpyruvylshikimate-3-phosphate (EPSP) synthase catalyzes the addition of the enolpyruvyl moiety of phosphoenolpyruvate (PEP) to shikimate-3-phosphate (S-3-P) (Fig. 1). This reaction constitutes one step of the shikimate pathway. Products of the shikimate pathway are necessary precursors for the synthesis of aromatic amino acids, and other aromatic compounds. EPSP synthase is an essential enzyme in organisms such as plants, bacteria, and fungi which must synthesize aromatic amino acids *de novo*. In plants the enzyme is primarily localized in the plastids which are the major sites of synthesis of aromatic compounds. EPSP synthase is not found in animals.

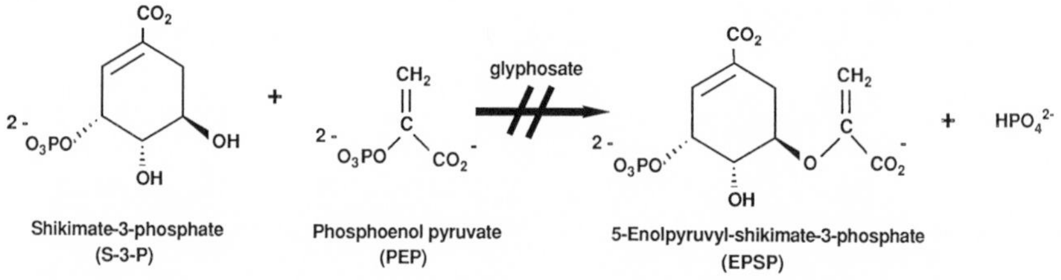

Shikimate-3-phosphate (S-3-P) + Phosphoenol pyruvate (PEP) —glyphosate→ 5-Enolpyruvyl-shikimate-3-phosphate (EPSP) + HPO_4^{2-}

Fig. 1. EPSP synthase catalyzes the condensation of phosphoenol pyruvate and shikimate-3-phosphate to form EPSP and inorganic phosphate. The reaction is inhibited by glyphosate.

EPSP synthase is of special agronomic interest since it is the specific target of glyphosate (N-phosphonomethyl glycine), a widely used broad spectrum herbicide. Because EPSP synthase is essential to the growth of all plants glyphosate is currently used primarily for field sterilization and selective application. Several features make glyphosate an especially useful herbicide. Glyphosate is rapidly taken up by the leaves of plants and is rapidly translocated to other organs. Plants are incapable of significant metabolism of glyphosate. Since the target enzyme for glyphosate is absent from all animals it shows very low toxicity, similar to that of table salt. Glyphosate is rapidly degraded by microorganisms in the soil eliminating residue problems. Because of this feature crops can be planted in a field a very short time after weeds are eliminated with glyphosate.

The utility of glyphosate could be significantly enhanced if glyphosate resistant crop plants could be produced. It could then be used as a safe, highly effective, selective herbicide. One of the goals of the Plant Molecular Biology Group at Monsanto has been to produce such plants. During the course of this project we have studied the structure and regulation of EPSP synthase genes of higher plants. We have developed an in vitro system to investigate the transport of EPSP synthase into chloroplasts, and have produced large quantities of plant EPSP synthase in Escherichia coli to aid in biochemical characterization of the enzyme. In this paper we report on our findings concerning this essential gene, and on our progress in engineering glyphosate resistance in higher plants.

ISOLATION OF CLONES FOR PLANT EPSP SYNTHASE

Due to the specific subcellular localization of EPSP synthase in plants we reasoned that we could most effectively engineer glyphosate resistant plants using an EPSP synthase gene derived from a plant. To facilitate isolation of the petunia gene a cell line (MP4-G) of *Petunia hybrida* resistant to glyphosate was developed by stepwise selection in increasing concentrations of glyphosate. Subsequent study of this cell line[1] showed that the resistance was due to an increased level of EPSP synthase. The sequence of the amino terminus of EPSP synthase from these cells allowed us to design a degenerate oligonucleotide probe for isolation of a cDNA to petunia EPSP synthase. A cDNA library was made from RNA of the resistant cell line, and was screened with the synthetic probe. Several rounds of screening led to the isolation of a full length clone. A map of the full length petunia EPSP synthase cDNA is shown in Figure 2. The amino acid sequence predicted from the sequence of the cDNA clone showed that 72 amino acid residues were present in the primary translation product that were not found in the mature protein as determined by protein sequencing.[2] These 72 amino acids represent the transit peptide which is responsible for directing the enzyme to the plastids.[3] Once inside the plastid the transit peptide is removed to form the mature enzyme which contains 444 amino acids. The molecular weight calculated from the deduced amino acid sequence (47.6 kd) is in close agreement with the molecular weight as measured by gel electrophoresis (49-55 kd).[3]

The petunia cDNA was used as a probe to isolate a cDNA to tomato EPSP synthase from a library constructed from RNA from tomato pistils. A map of the tomato cDNA is shown in Figure 2. The region of the tomato cDNA encoding the mature protein is very similar to the corresponding coding region of the petunia cDNA, the nucleotide sequences being 87% identical. The deduced amino acid sequence for mature tomato EPSP synthase shows that the enzyme is identical in size to the petunia enzyme in consisting of 444 amino acids. The amino acid sequence is identical to that of petunia in

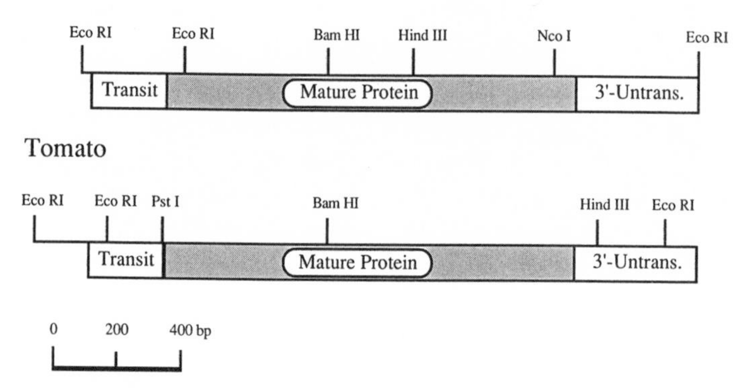

Fig. 2. Structure of the cDNA clones for petunia and tomato EPSP synthase.

93% of the residues. The tomato EPSP synthase transit peptide is 76 amino acids in length, 4 amino acids longer than that of the petunia enzyme. The degree of conservation in this region is much lower, with only 58% of the residues being identical between the two plants. This difference in rate of divergence between the two regions of the protein indicates that the functional constraints on the transit peptide are much looser than those on the enzymatically active region.

To isolate a genomic clone for petunia EPSP synthase, a genomic library was constructed from DNA isolated from the glyphosate resistant petunia cell line, MP4-G. Several clones were isolated from the library by screening with cloned petunia EPSP synthase fragments, and one of these was found to contain a complete gene. The regions of the genomic clone that showed hybridization to the cDNA were subcloned and sequenced. The sequence showed that the gene contained seven introns (Fig. 3). Two of the introns are large, and the gene spans nine kilobases. The sequence of the exons was found to be identical to the sequence of the cDNA. Hybridization of the cDNA to Southern blots of DNA from MP4-G cells and glyphosate sensitive MP4 cells (the progenitor to the glyphosate resistant line) showed that a gene amplification event had occured during the development of glyphosate resistance.[2] It is likely that

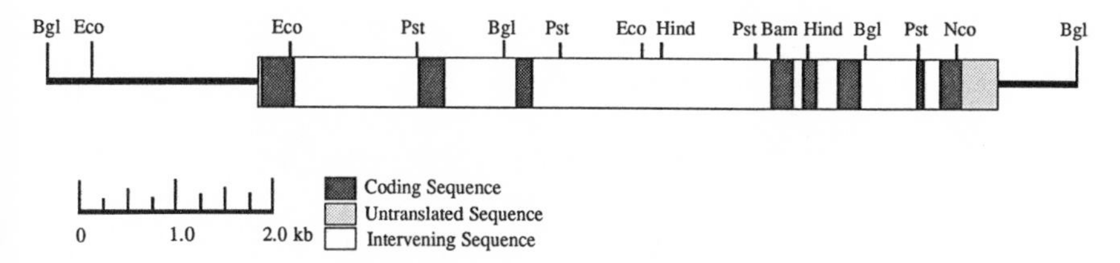

Fig. 3. Structure of a petunia EPSP synthase gene.

this gene amplification is responsible for the over-production of EPSP synthase in this cell line, as has been observed in other drug-resistant cell cultures.[4] The hybridization patterns also show that there exists a second set of unamplified sequences that hybridize to the EPSP synthase cDNA. These sequences are not amplified in the MP4-G cells. The restriction map of the gene that we have cloned corresponds to that of the amplified gene.

PATTERN OF EXPRESSION OF EPSP SYNTHASE

Although aromatic amino acids are required for growth and normal metabolism of all cells, it is possible that some organs or tissues of the plant have a relatively greater requirement for rapid synthesis of these compounds. Since products of the shikimate pathway are necessary precursors of other aromatic compounds such as lignin and pigments, there may be a greater need for flux through this pathway in cells where these compounds are synthesized. It is, therefore, possible that the level of EPSP synthase, and EPSP synthase mRNA, may vary in the different organs of the plant. To test this hypothesis RNA was purified from leaves at three stages of development, and from mature flowers of petunia plants. The RNA was used to produce a Northern blot. The blot, which was probed with labelled petunia EPSP synthase cDNA and two control probes, is shown in Figure 4. The amount of EPSP synthase mRNA is much greater in the flowers than in any stage of development of the leaves. In leaves, the EPSP synthase message is only visible in long exposures (data not shown). Further experiments showed that the level of expression in the leaves is typical of all of the vegetative organs. Subdivision of the flowers into pistils, anthers and petals showed

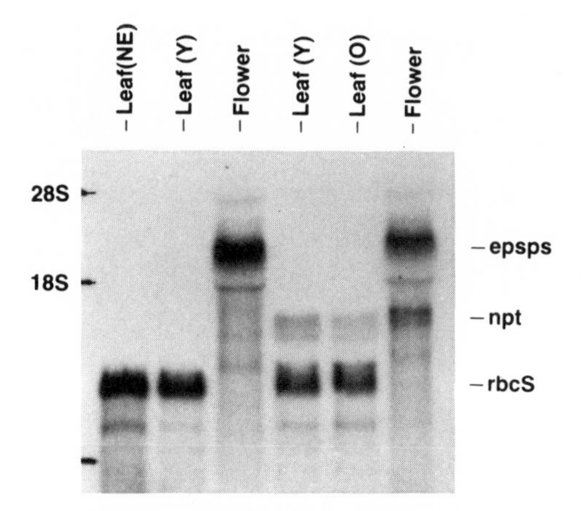

Fig. 4. Northern blot comparing the levels of expression of EPSP synthase in leaves and flowers of petunia plants. Total RNA (40 μg) from the indicated organs was electrophoresed on a formaldehyde gel and transferred to a nylon filter. The filter was hybridized with radiolabelled probes for petunia EPSP synthase, neomycin phosphotransferase (npt, E. coli), and the small subunit of ribulose-bis-phosphate carboxylase (rbcS) of petunia. The first three lanes are RNA from a wild-type plant. The last three lanes are from a plant that had been transformed with a vector which included a chimeric npt gene that expresses in plants. 18S And 28S represent the migration of the 18S and 28S ribosomal RNAs; Leaf(NE), RNA from newly emerging leaves; Leaf(Y), RNA from young leaves; Leaf(O), RNA from mature leaves. The migration of the RNAs which hybridize to the three probes are indicated at the right of the figure.

that the pistils and anthers contain EPSP synthase RNA at a level similar to that observed in leaves, while the petals exhibit a level that is 20-50-fold higher.[14] Similar experiments were performed on RNA purified from isolated tomato organs. In tomato it was found that the steady-state level of EPSP synthase mRNA varies at most two-fold between the different organs.[14] The level is slightly higher than the level in petunia leaves, but is much lower than that found in petunia petals.

We have yet to formulate an adequate explanation for the dramatic difference in the pattern of expression of EPSP synthase between petunia and tomato. Tomato and petunia are closely related, both being members of the family Solanaceae. The only obvious feature of petunia petals that differentiates them from those of tomato is that in many petunia cultivars the petals are bright purple. The purple pigments are aromatic compounds which are synthesized in the petals from products of the shikimate pathway. However, the actual quantity of pigment in the petals does not seem sufficient to require the level of over-expression of EPSP synthase seen in the Northern blots. The flavonoid pigments responsible for the yellow color of tomato petals are also derived from the shikimate pathway, yet there is no increase in the level of EPSP synthase mRNA in tomato flowers. At present we are examining the pattern of expression of EPSP synthase in other higher plants to see if the petunia expression pattern is found in any other species.

COMPARISONS OF EPSP SYNTHASES FROM DIFFERENT ORGANISMS

EPSP synthase genes from bacteria[5,6] and fungi[7] have been isolated and sequenced. In the fungus *Aspergillus nidulans* the EPSP synthase activity is associated with a large multifunctional protein, the AROM complex.[7] This complex carries out five steps of the shikimate pathway. A region of the protein shows substantial homology to the EPSP synthase of *E. coli* and is presumed to be the region responsible for the EPSP synthase activity.[7] We have compared the protein sequences of our plant EPSP synthase to those of *E. coli*, and *Aspergillus* and used the degree of homology to construct a phylogenetic tree (Fig. 5). The pattern of relatedness is surprising. The plant genes are much more closely related to the *E. coli* sequence than to that of the other eukaryote, *Aspergillus*. In *Aspergillus* the AROM complex is localized in the cytoplasm, while in plants the EPSP synthase appears to accumulate only in the plastids. It has been hypothesized that plastids are derived from prokaryotic endosymbionts.[8,9] If this is the case then it is possible that the EPSP synthase gene that we have characterized is a gene that, at some time during the evolution of higher plants, has migrated from the plastid to the nucleus. Such a gene would be expected to share homology with genes of prokaryotes. We consider this

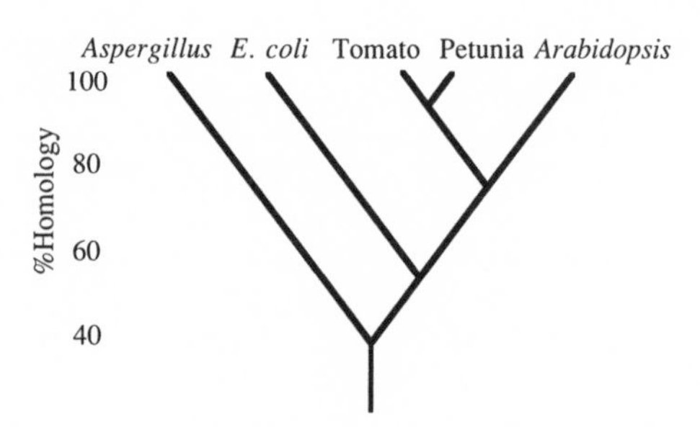

Fig. 5. Phylogenetic tree based on the number of amino acids in the EPSP synthase of the organism listed that are identical. The petunia and tomato sequences are from Reference 14, the E. coli sequence is from Reference 5, and the Aspergillus sequence is from Reference 7.

the most likely explanation for the structure of the phylogenetic tree (Fig. 5).

TRANSLOCATION OF EPSP SYNTHASE INTO CHLOROPLASTS

Although plastids have an endogenous genome it is now known that the majority of proteins localized in the chloroplast are encoded by nuclear genes.[10] The nuclear genes encode precursor proteins that are translated on free cytoplasmic ribosomes prior to transport into the chloroplasts. The precursor proteins include amino-terminal extensions termed "transit peptides" which are responsible for the post-translational uptake of the proteins by the plastids. After uptake, the transit peptide is removed by a sequence-specific protease to form the mature protein.

We have developed an in vitro system to study the uptake and processing of EPSP synthase[3] (Fig. 6). For these experiments the full length petunia EPSP synthase cDNA was inserted into a plasmid vector which included a promoter for a phage RNA polymerase. This plasmid can be used to synthesize substantial quantities of EPSP synthase mRNA in vitro using a purified phage RNA polymerase that is commercially available. In vitro translation of this RNA in the presence of radioactively labelled amino acids is then performed to produce labelled precursor protein.

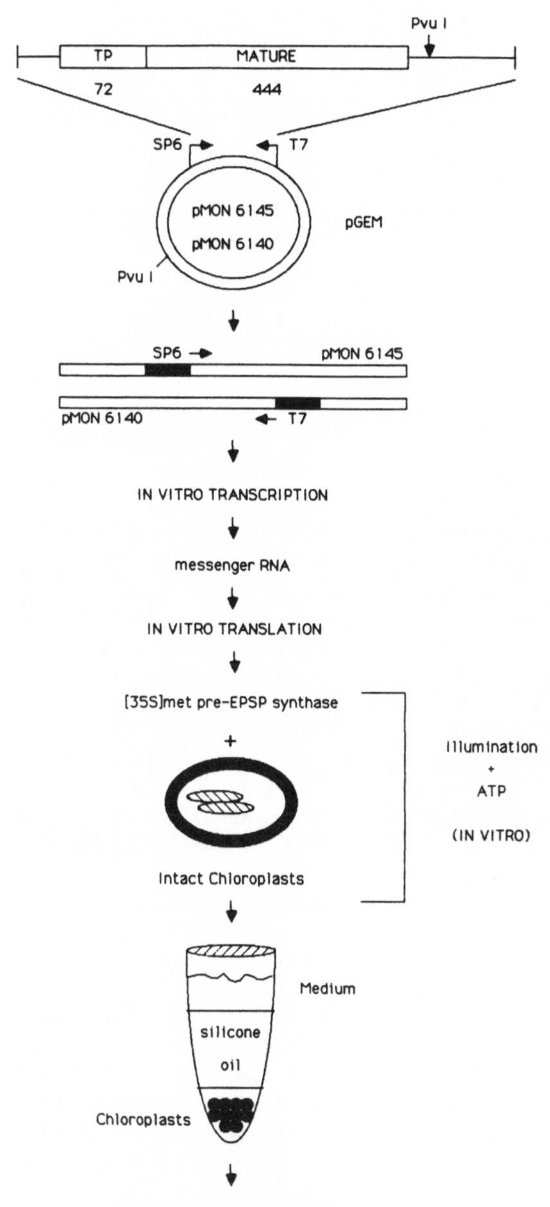

Fig. 6. Method of generating EPSP synthase precursor protein and assaying its uptake _in vitro_.

When the precursor protein is incubated with isolated chloroplasts from lettuce (Latuca sativa) it is rapidly taken up and processed to the mature form.[3] We also demonstrated that when the petunia transit peptide is attached to the amino terminus of E. coli EPSP synthase it can efficiently direct this enzyme to the chloroplast in vitro.[11] This observation may be useful in genetically engineering herbicide resistance or other desirable traits in plants (see below).

The in vitro transcription/translation system provides sufficient protein for biochemical analysis. We have found that the pre-EPSP synthase of petunia is catalytically active.[3] This implies that the active site regions of pre-EPSP synthase are capable of folding into the configuration found in the mature enzyme prior to uptake and processing. We have additionally shown that the pre-EPSP synthase is inhibited by glyphosate with approximately the same concentration dependence as the mature protein.[3]

Our finding that glyphosate binds to and inhibits the catalytic activity of the precursor to petunia EPSP synthase led us to study the effect of glyphosate on transport into the chloroplast. We find that including glyphosate and shikimate-3-phosphate in the incubation medium dramatically inhibits the ability of isolated chloroplasts to import and process EPSP synthase (Fig. 7). Shikimate-3-phosphate alone has no effect (data not shown). The same experiment was performed using precursor to a glyphosate resistant mutant form of EPSP synthase. This enzyme is resistant by virtue of an increase in the K_i for glyphosate. The uptake of the mutant EPSP synthase is only slightly inhibited by glyphosate (Fig. 7). This demonstrates that it is the binding of glyphosate to the pre-enzyme that is responsible for the decreased rate of uptake, rather than some secondary effect of glyphosate in the medium. It is possible, therefore, that the phytotoxic effect of glyphosate is a combination of direct inhibition of EPSP synthase activity and a decrease in the amount of enzyme in the plastids due to inhibition of enzyme transport.

CHARACTERIZATION OF MATURE EPSP SYNTHASE

EPSP synthase is produced in extremely small quantities in plants.[1] To facilitate the characterization of the

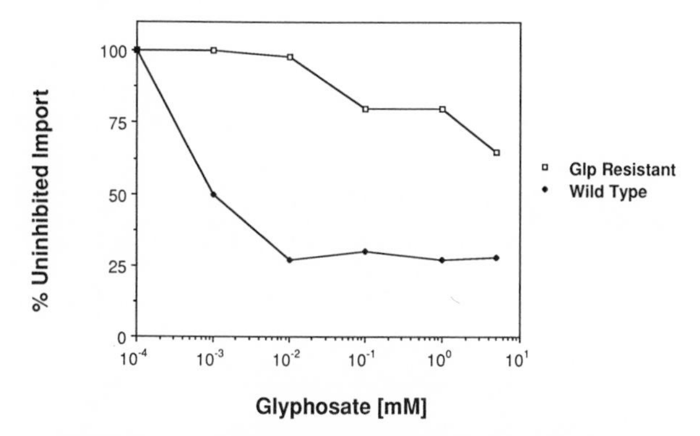

Fig. 7. Glyphosate inhibition of uptake of EPSP synthase by isolated chloroplasts. Uptake and processing of pre-EPSP synthase was measured in the presence of increasing concentrations of glyphosate as outlined in Figure 6. Independent experiments were performed on a wild type pre-enzyme, and on a mutant pre-enzyme which included a coding sequence with reduced affinity for glyphosate.

petunia enzyme we have constructed a vector for production of large amounts of the mature form of petunia EPSP synthase in *E. coli*.[12] This enzyme has been shown to be identical to the enzyme isolated from petunia cells in terms of the K_M's for PEP and S-3-P, and the K_i for glyphosate. We are currently performing chemical modification and inhibitor studies on enzyme purified from the over-producing bacterial cells in an effort to define the regions of the enzyme that are critical to catalytic activity and glyphosate binding.

We have used both *in vitro* and *in vivo* methods to isolate a number of different glyphosate tolerant forms of both plant and bacterial EPSP synthases. These mutants exhibit a broad range of resistance as shown in Figure 8. More detailed analysis of the nature of these mutations will provide further insight into the amino acids involved in catalysis and glyphosate binding.

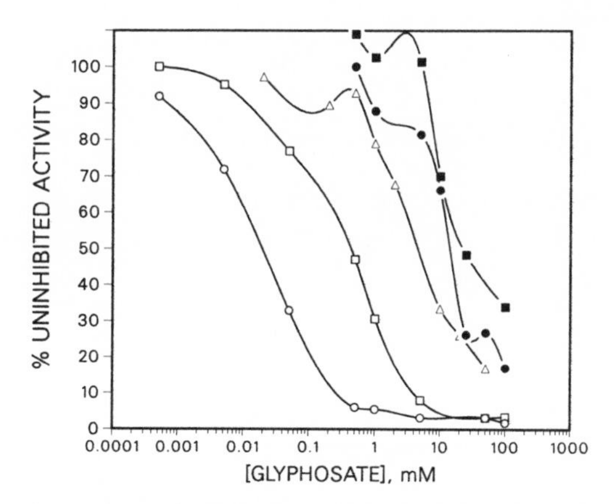

Fig. 8. Glyphosate inhibition of enzyme activity for wild type and mutant enzymes. EPSP synthase was measured by incorporation of radiolabelled PEP into EPSP[12] in increasing concentrations of glyphosate. The curve with open circle data points represents the activity of wild type enzyme while the other curves represent the activities of mutant enzymes that we have isolated. The most resistant mutant has an I_{50} value 1000-fold greater than the wild type enzyme.

ENGINEERING GLYPHOSATE TOLERANCE IN PLANTS

We have used the technique of *Agrobacterium* mediated gene transfer to engineer plants with enhanced tolerance to glyphosate. In our initial experiments the wild type petunia EPSP synthase cDNA was placed under control of the promoter for the 35S transcript of cauliflower mosaic virus (CaMV). This promoter leads to high-level expression of attached genes in transgenic plants.[13] Introduction of the chimeric 35S/EPSP synthase genes into petunia cells led to the formation of callus that was able to grow on concentrations of glyphosate sufficient to completely inhibit proliferation of wild type callus.[2] Transformed petunia plants were regenerated from similar transformation experiments. These plants were shown to be tolerant to a dose of formulated glyphosate (Roundup® herbicide) equivalent to 0.8 lb/acre, approximately four time the quantity necessary to kill 100% of the control plants transformed with a vector that did not contain the chimeric EPSP synthase gene.

Analysis of tissues transformed with the chimeric EPSP synthase gene showed that they contained ∿20-fold more EPSP synthase that the wild type controls.[2] Subsequent experiments have shown that this additional enzyme is localized in the plastids (data not shown). This demonstrates that overproduction of EPSP synthase confers glyphosate resistance on plants and plant cell cultures. Similar resistance due to overproduction of the target enzyme of an inhibitor has been previously observed in a number of cultured cell systems.[1,4]

To achieve higher levels of glyphosate resistance we have produced a number of glyphosate resistant forms of plant and bacterial EPSP synthase either by *in vivo* or *in vitro* mutagenesis (Fig. 8). In the case of the resistant plant enzymes the presence of the transit peptide ensures proper transport of the glyphosate resistant enzyme to the chloroplasts where it will most effectively interact with the other enzymes of the shikimate pathway. As shown above, attachment of an appropriate transit peptide sequence to the amino terminus of a bacterial enzyme will also direct proper uptake into the plastids. Using these methods high levels of tolerance to Roundup® have been achieved in petunia, tobacco and tomato plants.

CONCLUSIONS

We have isolated EPSP synthase cDNA clones from tomato, and cDNA and genomic clones from petunia. The genes are found to be very similar to the *E. coli* EPSP synthase gene in the sequence for the mature protein. The plant enzymes differ from those of bacteria in that they are synthesized as precursors which include transit peptides responsible for translocating the enzyme to the chloroplast. In petunia the gene is found to express at a much higher level in petals than in other organs of the plant. In contrast the level of expression is constant in different organs of tomato plants.

Production of plant EPSP synthase precursor *in vitro* has facilitated the study of uptake of the enzyme by chloroplasts. Using these methods we have shown that a bacterial enzyme is efficiently translocated into chloroplasts when the plant transit peptide is attached to the amino terminus of the protein. We have constructed bacterial expression vectors that direct the synthesis of large amounts of plant EPSP synthase in *E. coli*. Enzyme purified from the overproducing bacteria is being used in inhibitor and protein modification studies.

Plants which tolerate high levels of glyphosate application have been engineered by transformation with vectors which lead to overproduction of wild-type EPSP synthase, or production of mutant enzymes with an altered affinity for glyphosate. These studies promise to greatly extend the utility of this safe effective herbicide by allowing it to act as a selective herbicide. We are continuing our research on the structure and expression of EPSP synthase genes by examining other species. We are currently working to produce effective glyphosate tolerance in additional species of crop plants.

ACKNOWLEDGMENTS

We thank Alan Smith for his critical reading of the manuscript, and Ernest Jaworski for his continual support of this work.

REFERENCES

1. STEINRÜCKEN, H.D., A. SCHULL, N. AMRHEIN, C.S. PORTER, R.T. FRALEY. 1986. Overproduction of 5-enolpyruvylshikimate-3-phosphate synthase in a glyphosate-tolerant *Petunia hybrida* cell line. Arch. Biochem. Biophys. 244: 169-178.
2. SHAH, D.M., R.B. HORSCH, H.J. KLEE, G.M. KISHORE, J.A. WINTER, N.E. TUMER, C.M. HIRONAKA, P.R. SANDERS, C.S. GASSER, S. AYKENT, N.R. SIEGEL, S.G. ROGERS, R.T. FRALEY. 1986. Engineering herbicide tolerance in transgenic plants. Science 23: 478-481.
3. DELLA-CIOPPA, G., C. BAUER, B.K. KLEIN, D.M. SHAH, R.T. FRALEY, G.M. KISHORE. 1986. Translocation of the precursor of 5-enolpyruvylshikimate-3-phosphate synthase into chloroplasts of higher plants *in vitro*. Proc. Natl. Acad. Sci. U.S.A. 83: 6873-6877.
4. SCHIMKE, R.T. 1984. Gene amplification in cultured cells. Cell 37: 705-713.

5. DUNCAN, K., A. LEWENDON, J.R. COGGINS. 1984. The complete amino acid sequence of Escherichia coli 5-enolpyruvylshikimate-3-phosphate synthase. FEBS Lett. 170: 59-63.
6. STALKER, D.M., W.R. HIATT, L. COMAI. 1985. A single amino acid substitution in the enzyme 5-enolpyruvylshikimate-3-phosphate synthase confers resistance to the herbicide glyphosate. J. Biol. Chem. 260: 4724-4728.
7. CHARLES, G., J.W. KEYTE, W.J. BRAMMAR, M. SMITH, A.R. HAWKINS. 1986. The isolation and nucleotide sequence of the complex AROM locus of Aspergillus nidulans. Nucleic Acids Res. 14: 2201-2213.
8. MARGULIS, L. 1970. Origin of Eukaryotic Cells. Yale University Press, New Haven, Connecticut, 349 pp.
9. WEEDEN, N.F. 1981. Genetic and biochemical implications of the endosymbiotic theory of the origin of the chloroplast. J. Mol. Evol. 17: 133-139.
10. CASHMORE, A., L. ZABO, M. TIMKO, A. KAUSCH, G. VAN DEN BROECK, P. SCHREIER, H. BOHNERT, L. HERRERA-ESTRELLA, M. VAN MONTAGUE, J. SCHELL. 1985. Import of polypeptides into chloroplasts. Bio/Technology 3: 803-808.
11. DELLA-CIOPPA, G., S.C. BAUER, M.L. TAYLOR, D.E. ROCHESTER, B.K. KLEIN, D.M. SHAH, R.T. FRALEY, G.M. KISHORE. 1987. Targeting of a herbicide-resistant enzyme from Escherichia coli to chloroplasts of higher plants. Bio/Technology 5: 579-584.
12. PADGETTE, S.R., Q.K. HUYNH, J. BORGMEYER, D.M. SHAH, L.A. BRAND, D.B. RE, B.F. BISHOP, S.G. ROGERS, R.T. FRALEY, G.M. KISHORE. 1987. Bacterial expression and isolation of Petunia hybrida 5-enolpyruvylshikimate-3-phosphate synthase. Arch. Biochem. Biophys. 258: 564-573.
13. SANDERS, P.R., J.A. WINTER, S.G. ROGERS, R.T. FRALEY. 1987. Comparison of cauliflower mosaic virus 35S and nopaline synthase promoters in transgenic plants. Nucleic Acids Res. 15: 1543-1558.
14. GASSER, C.S., J.A. WINTER, C.M. HIRONAKA, D.M. SHAH. 1988. Structure, expression and evolution of the 5-enolpyruvyl-shikimate-3-phosphate synthase genes of petunia and tomato. J. Biol. Chem. (in press).

Chapter Four

MOLECULAR APPROACHES TO UNDERSTANDING CELLULAR RECOGNITION IN PLANTS

ANTONY BACIC AND ADRIENNE E. CLARKE

Plant Cell Biology Research Centre
School of Botany
The University of Melbourne
Parkville
Victoria 3052
Australia

INTRODUCTION

Plant cells, like animal cells, have the capacity to discriminate between self and non-self. Examples of interactions involving cell-cell recognition in plants are those between pollen and pistils during fertilization, host and microorganism and somatic cells of plants during grafting (for reviews [1,2]). In this paper, we describe approaches we have used to understand the recognition events which occur during fertilization. In 1980 we presented a review of our efforts to understand this subject at the Phytochemical Society Meeting held at Washington State University;[3] we now present a summary of the progress we and others have made in the intervening seven years.

BIOLOGY AND GENETICS OF FERTILIZATION

This subject has been reviewed comprehensively many times. Here we present an outline of the subject, to provide a framework for understanding the rationale of the experimental work described. For a more detailed treatment of the subject, the reader is referred to the excellent reviews by de Nettancourt,[4] Lewis,[5] Heslop-Harrison,[6] Gibbs,[7] Pandey,[8] Lawrence[9] and Linskens.[10]

The interacting cells during pollination are the pollen grain or pollen tube and the tissues of the female pistil. During a compatible mating, pollen which is carried to the pistil by wind, water or animal vectors hydrates and germinates to produce a pollen tube. The pollen tube carries the two sperm cells, the vegetative cell and the cytoplasm in the tip as it grows through the stigma and style to the ovary, where fertilization occurs (Fig. 1). The two sperm cells are released in the embryo sac, one to fertilize the ovum and the other to fuse with the polar nuclei to form the primary endosperm nucleus.

The female pistil is able to select compatible pollen from the range of pollen it may be presented with; thus intergeneric and interspecific pollen is usually prevented from fertilizing the ovum. Intraspecific crosses are often successful; however, in many plant families, self-incompatibility genes (*S*-genes) operate to prevent self-fertilization. The *S*-genes often, but not always, have a single locus with many alleles. For example, in *Papaver rhoeas* there are forty or more alleles.[11] This is effectively a recognition and rejection of "self" and provides a relatively simple system for studying the mechanism of the recognition events involved.

There are two homomorphic self-incompatibility systems: those with gametophytic control and those with sporophytic control. Sporophytic self-incompatibility is apparently restricted to two families, the Crucifereae and the Compositae. In this system, the *S* genotype of the pollen producing plant determines the incompatibility reaction of the pollen. Thus, if an allele of the *S* gene in the pollen parent is also present in the female tissues, pollen tube growth is inhibited at the stigma surface (Fig. 2). The arrest of growth is characteristically accompanied by formation of callose deposits at places

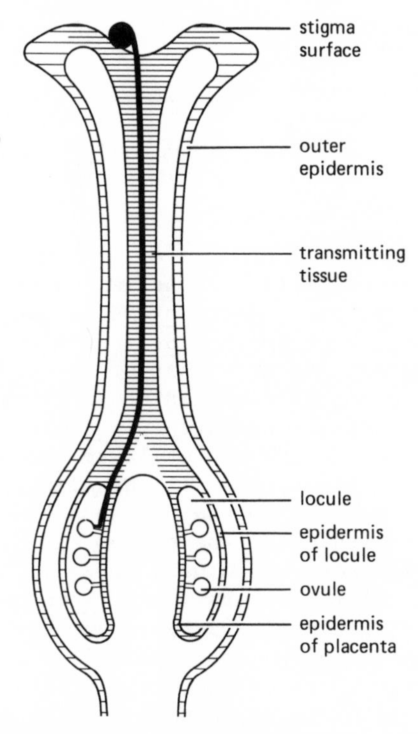

Fig. 1. Diagram of compatible pollination of a mature pistil of Nicotiana alata. The secretory epidermis of the stigma, transmitting tract of the style and the epidermis of the placenta are indicated by close cross-hatching; these are the tissues which hybridize with $\underline{S}_2$ cDNA and correspond to the path taken by the pollen tube before entering the ovule (from Reference 51).

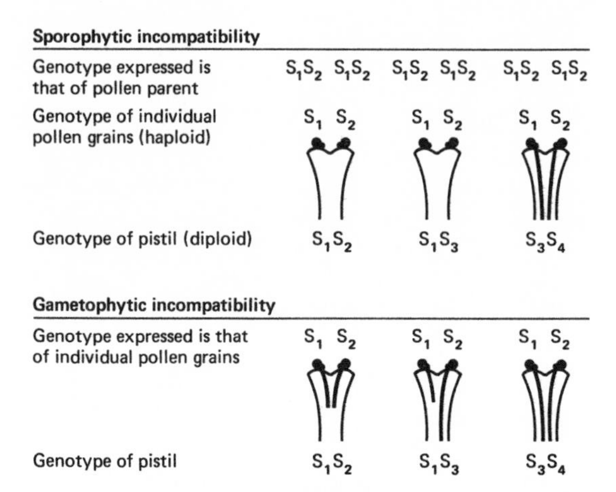

Fig. 2. Behavior of pollen in the two major self-incompatibility systems. Sporophytic incompatibility: The pollen parent genotype is S_1S_2. When an allele in the pollen parent is matched with that of the pistil (e.g., S_1S_2 or S_1S_3), pollen germination is arrested at the stigma surface. Where there is no match (S_3S_4) the pollen may germinate and grow through the style to the embryo sac. The central panel only applies if the allele S_1 is dominant to or codominant with S_2 in the pollen and S_1 is dominant to or codominant with S_3 in the style. If S_3 is dominant to S_1 in the style, or S_2 dominant to S_1 in the pollen, the pollen from S_1S_2 will be compatible. Gametophytic incompatibility: The pollen parent genotype is S_1S_2. When an allele in the individual haploid pollen grain is matched with either allele in the diploid style tissues, growth of the pollen tube is arrested, usually in the style. For example, both S_1 and S_2 pollen are inhibited in a S_1S_2 style but, unlike the sporophytic system, the S_2 pollen will grow successfully through S_1S_3 style. Where there is no match of alleles (e.g., S_1S_2 pollen grains on a S_3S_4 pistil) the pollen tubes of both genotypes grow through the style to the embryo sac (from Reference 2).

on the stigma surface which make contact with the pollen and also in the pollen grain or rudimentary pollen tube. The recognition event is then between the surfaces of the stigmatic papillae and those of the pollen grains or rudimentary pollen tubes, and must involve information encoded by the S-alleles in the pollen and stigma.

The other type of homomorphic self-incompatibility is that with gametophytic control. This is the commonest type of self-incompatibility, and is characterized by production of pollen which expresses its own genotype. A pollen tube carrying a single given allele is inhibited if the same allele is present in the diploid tissues of the style (Fig. 2). The recognition event is therefore between the pollen grain or the pollen tube and the tissues of the pistil and must involve molecules encoded by the S-gene in both partners.

Characteristically in the gametophytic system, both incompatible and compatible pollen grains hydrate and germinate normally. Both types of tube penetrate the stigma and enter the style but at some point during growth through the style, the tip of the incompatible pollen tube becomes abnormal and may swell and burst. The compatible and incompatible tubes are morphologically distinct. The compatible tubes stain with the aniline blue fluorochrome to show regular deposits of callose within the tube. These deposits are believed to effectively cut off the growing tip from the empty pollen grain. The incompatible tubes have irregular deposits of callose, fluoresce more strongly over the whole wall after staining with the aniline blue fluorochrome and have characteristically thickened walls. The changed appearance of the incompatible tubes and the ultimate cessation of tube growth may be the result of a recognition event at the stigma surface which has a slow effect on tube growth. Alternatively, there may be an interaction between molecules of the style tissues and the pollen tube growing through the style which leads to the arrest of tube growth. The position within the female tissues at which growth of the pollen tube is arrested is often within the style tissue, for example, in members of the Solanaceae such as *Nicotiana alata*, *Lycopersicon peruvianum* and *Petunia hybrida* and members of the Rosaceae such as *Prunus avium*. However, some other gametophytic self-incompatibility systems such as those of *Secale cereale* and *Oenothera organensis* are different in that the tubes are arrested either on the stigma surface or just after penetration into the style.[12,13] In others such as *Beta vulgaris*[14] pollen tubes may reach the ovary before their growth is arrested. Thus, while we might anticipate an underlying common mechanism involving the *S*-gene products in pollen and pistil for all the gametophytically connected self-incompatibility systems, it may be modified to result in the very different manifestations in different families. Notwithstanding the limitations of attempting to generalize from studies on one or a few systems, we have taken the view that by understanding one system in detail, we will build a knowledge base that will lead to a more general understanding of self-incompatibility.

SELF-INCOMPATIBILITY IN *NICOTIANA ALATA* (SOLANACEAE)

Since 1980 we have focussed our efforts to understand the molecular events of self-incompatibility in *N. alata*, an ornamental tobacco. This plant has a homomorphic gametophytic self-incompatibility system which is controlled by a single *S*-gene with multiple alleles.[2,15,16] In a compatible pollination, the pollen germinates to produce a tube which grows through the central transmitting tissue of the style to the ovary. The tube is in contact with the extracellular mucilage of the transmitting tissue during its growth through the style. The pathway is always intercellular as the pollen tubes grow between the files of transmitting tissue cells (Fig. 3). In a self-mating, the pollen germinates and produces a tube which grows into the upper part of the style where growth is arrested. We have studied three aspects of self-incompatibility in *N. alata* in order to understand the recognition events leading to the arrest of self-pollen tubes:

1. *Pollen tube structure*. Before we could devise experiments aimed at understanding the biosynthesis of pollen tubes and the fault which develops in self-incompatible tubes, we had to understand the nature of the wall components.

2. *The style mucilage components*. The pollen tubes are in contact with the extracellular mucilage of the style transmitting tissue on their pathway to the ovary. As proteoglycans of the class of arabinogalactan-proteins are major components of this mucilage, we decided to establish their structural features and then attempt to understand their function.

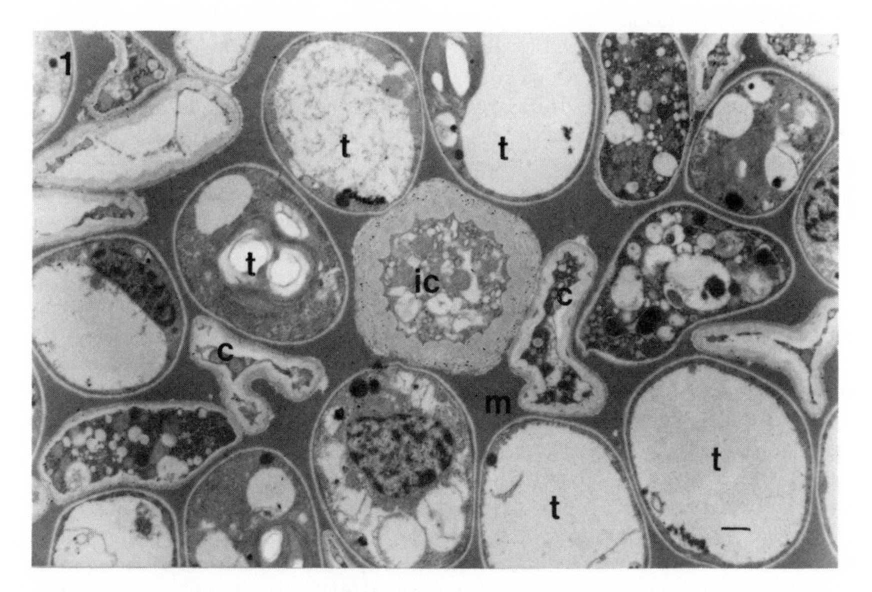

Fig. 3. Transverse section of the transmitting tissue of a style of Nicotiana alata fixed 12 h after pollination with self-compatible pollen. The section was taken 4-7 mm below the stigma surface and shows transmitting tissue cells (t), intercellular matrix (m), compatible (c) and incompatible (ic) pollen tubes. It was immuno-gold labelled and then stained with uranyl acetate and lead citrate. Bar = 1 μm. (From Reference 22.)

3. The genetic control of self-incompatibility. The most significant advances in plant biology in the last seven years have come from the application of recombinant DNA techniques. The application of this technology has given us detailed knowledge of the S-gene-associated molecules in the female tissues, and promises to lead us to an understanding of the allelic specificity.

We now describe in detail the advances in knowledge we have made in these three areas in the last seven years.

Pollen-tube Walls

The fluorescence of pollen tubes of many species after staining with decolorized aniline blue indicates that their walls contain a β 1→3-glucan. The specificity of this staining procedure has been defined, as the fluorochrome which occurs in aniline blue has been identified, synthesized and the specificity of the synthetic fluorochrome for a number of glucans of defined linkage types studied.[17,18] The primary specificity is for β 1→3-D-glucans, although some β 1→3, 1→4-D-glucans also bind the fluorochrome. Thus, the fluorescent staining induced by this fluorochrome in the cell walls and plugs in pollen tubes of many species is indicative of their content of 1→3 linked glucosyl residues but does not exclude the possibility of some 1→4 linkages within the glucan chain. Cell walls of pollen tubes of *N. alata* grown *in vitro*, prepared by a mild method which involved only breakage of the cells, removal of the cytoplasm by water washing and drying of the wall preparation with solvents, had glucose and arabinose as the major monosaccharides[19] (Table 1). Monosaccharide analysis of *in vitro* grown pollen tubes of six other species (for review[1]) show that glucose is the major component in each case, with either arabinose or galactose being the next most abundant monosaccharide. Linkage analysis showed that the main linkage type was 3-linked glucose, but that there was also smaller amounts of 3,6 linked and terminal residues. The analyses are consistent with a 1→3-glucan, branched through C(O)6. Some 4-linked glucosyl residues were detected, and these probably reflect the cellulose content of the tube wall, although it is possible that some may be present in a mixed linkage 1→3,1→4-glucan chain. The other major monosaccharide was arabinose, primarily 5-linked with a low proportion of 2,5-linked and terminal residues indicating the presence of a 1→5-linked arabinan with branching through C(O)2.[19] Arabinans of this general structure are also present in the pectic fraction of vegetative cell walls of a number of dicotyledons.[20] The pollen tube wall is bilayered, except at the tip. The arabinofuranosyl residues can be localized to the outer part of the wall by immunocytochemistry using a monoclonal antibody directed to terminal α-L-arabinofuranosyl residues[21] and by the finding that the outer wall is periodate sensitive.[22] The inner wall has an appearance typical of β 1→3-glucans in that it is electron lucent and not vulnerable to periodate oxidation in the Thiery test.[22] These analyses, although

Table 1. Neutral-monosaccharide composition of cell walls of germinated pollen grains of *Nicotiana alata*.

Monosaccharide	Amount (mol %)	
	Sulphuric-acid Hydrolysis[a]	Trifluoroacetic-acid Hydrolysis[b]
Rhamnose	0.9	1.2
Fucose	tr	tr
Arabinose	15.4	26.4
Xylose	1.0	1.2
Mannose	1.5	0.9
Galactose	3.3	4.0
Glucose	78.0	66.3

[a] Mean of determinations of three hydrolysates.

[b] Average of determinations on two hydrolysates.

tr = Present in trace amounts.

they tell us the major structural features of the wall, have limitations. For example, we are analysing tubes grown *in vitro* - what is the corresponding analysis for compatible and incompatible tubes grown *in vivo*? Indeed, how could this information be obtained when the tubes grow within a central solid transmitting tissue of the style? What is the composition of the wall at the growing tip which lacks the β 1→3-glucan? What materials are produced by the tube and are secreted through the wall?

These questions lead to a major question which must be addressed if we are eventually to understand self-incompatibility: how is the synthesis and deposition of the various wall components controlled? These problems can be tackled but there are significant gaps in our general knowledge base and technology which indicate that getting the next order of understanding will not be straightforward. For example, the fundamental knowledge of how cellulose and β 1→3-glucans are synthesized, by growing somatic plant cells, is not established.[23] We are now initiating a study of the biosynthesis of pollen tube wall components to establish a base for examining how the S-gene products act to cause aberrant tube growth and ultimate arrest of incompatible tubes.

Style Mucilage Proteoglycans: The Arabinogalactan-Proteins

Our early studies showed that the stigma surface of *Gladiolus* is covered at maturity by a sticky layer containing a high proportion of arabinogalatan-protein (AGP).[24] Labarca and Loewus[25] had examined another member of the Liliaceae, *Lilium longiflorum*, and found galactose and arabinose as major monosaccharides of the mucilage of both stigma and style. Although the fractions studied were probably mixtures of several carbohydrate-containing macromolecules, the data indicates that it is most likely that AGPs are major components of exudates of stigma and style of *L. longiflorum*.[26] We then surveyed pistils of a range of angiosperms and found that they all contained arabinogalactan proteins.[27] These proteoglycans are widespread in plant tissues and are characteristically produced on wounding.[28,29] As the AGPs are major buffer-soluble, high molecular weight components of styles, we undertook a detailed study of the AGPs in the female sexual tissues of *N. alata*. Our studies were facilitated by the knowledge that the red dyes known as Yariv reagents bind specifically to and precipitate AGPs. We used this dye as a cytochemical reagent to locate the AGPs as extra-cellular components of the transmitting tissue,[30] for quantification of AGPs in extracts,[31] and for separating different classes of AGPs on a charge basis by crossed-electrophoresis.[32] Using these methods, we established that in *N. alata* the AGPs of the stigma are developmentally regulated in that there is a 3 to 4 fold increase in their concentration from the stage of petal coloration to the mature flower. There is no corresponding increase in the concentrations of AGPs in the style. The charge nature of the classes of AGPs, in both stigma and style, are distinctive, and change during maturation.[33] The amount of extractable AGP per stigma increased after pollination. The increase was of the order of 40% over that extrated from unpollinated stigmas. The same increase was measured after pollination using either compatible pollen or incompatible pollen. There was, however, no change in the amount of extractable AGP in the style after pollination. Pollination with ethanol-vapor inactivated pollen also gave an increase in the amount of AGP in the stigma compared with the levels extracted from unpollinated styles, but the increase was less than half that measured after pollination with viable pollen.[33] The general patterns of crossed-electrophoresis of stigma and style extracts did not change after compatible or incompatible pollinations, or after pollination with alcohol-inactivated pollen. It is possible that the increases observed are a response to wounding of the stigma surface. The differential increase observed after pollination with viable or non-viable pollen may reflect the more extensive damage sustained by the stigma as it is penetrated by the germinated live pollen.

Another approach to the question of function of AGPs is to ask whether there is any relationship between *S* (self-incompatibility genotype) and the structure of the stigma AGPs. We isolated AGPs from *N. alata* stigmas, genotypes S_1S_3, S_2S_3, S_2S_2, and S_3S_3, by affinity chromatography on the J539-myeloma protein, which has a specificity for 1→6 linked β-D-galactopyranosyl residues, followed by gel permeation chromatography.[34] We found only minor differences between AGPs isolated from different genotypes. It seems unlikely that AGPs are directly involved in the expression of self-incompatibility. On the stigma surface, they may be responsible for generating conditions which are conducive for the adhesion and germination of pollen. In addition, they may provide nutrients for the growing pollen tube.[25,33]

S-Allele-associated Style Glycoproteins

Since Lewis' pioneering work on the immunological detection of antigenic material in pollen extracts of *O. organensis*,[35] which corresponded to particular *S*-alleles, there have been a number of reports of pistil proteins corresponding to *S*-alleles. Apart from our early work on proteins from *P. avium* styles,[36-38] there have been reports of *S*-allele-associated style glycoproteins in other gametophytically controlled systems: *Petunia* hybrida,[39] L. longiflorum,[40] L. peruvianum[41] and N. alata.[15] In the sporophytically controlled self-incompatibility systems of Brassica spp., a number of S-allele-associated stigma glycoproteins have also been identified.[42-45] The major advance in this area since 1980 has been the application of recombinant DNA and peptide sequencing techniques to obtain full nucleotide sequences for cDNAs encoding S-allele-associated glycoproteins from N. alata styles[46] and from stigmas of Brassica oleracea[47,48] and B. campestris.[49,50]

For N. alata, the cDNA encoding the 32K S_2 -allele-associated glycoprotein had a signal sequence typical of secreted eukaryotic proteins; the mature protein is hydrophilic in nature except for a relatively hydrophobic stretch of 15 amino acids at the N-terminus.[46] This N-terminal sequence is homologous with corresponding sequences from other S-allele-associated proteins from both N. alata and L. peruvianum.[41] This cDNA is a powerful probe for furthering our understanding of the mode of action of self-incompatibility. Hybridization of the probe to style sections shows that the gene is expressed, as would be expected, only in the transmitting tissue of the mature style. In the immature pistil, there is no expression in the style, but there is restricted expression in the stigma. Unexpectedly, there was strong expression in the ovary which was restricted to a single layer of cells, the epidermis of the placenta.[51] The expression in these cells marks precisely the pathway which pollen tubes take in their growth to the ovules. Why is it expressed in the ovary, when arrest of incompatible tubes takes place in the style? We can only put forward plausible but not definitive explanations. Perhaps it is as a "fail-safe" mechanism to ensure arrest of incompatible tubes if, for some reason such as environmental conditions, the growth of incompatible tubes progresses further than is usually allowed. This implies an active role for the style glycoprotein in ensuring arrest of the tube growth. On the other hand, it may be a positive guide for tube growth of compatible pollen, to ensure its growth to the correct target.

The cDNA encoding the style glycoprotein can also be used to detect other S-alleles. In all S-genotypes of N. alata examined, we detected, by northern analysis, homologous mRNA species of identical size (≅900 nucleotides);

the extent of hybridization was variable and in all cases significantly lower than the level obtained in styles bearing the $\underline{S}_2$-allele. This indicates either lower expression from these alleles, or a major sequence change, or a combination of both.[46,52] Southern analysis also shows variable and generally weak hybridization to all other alleles, pointing to major changes in sequence between different alleles (unpublished observations, R. Bernatzky, E. Cornish, S-L. Mau). The corresponding data for the sporophytic systems of B. oleracea[47,48] and B. campestris,[49,50] in contrast with N. alata show that there is strong homology between the stigma glycoproteins corresponding to different alleles in the two species. The N-terminal sequences of the S-allele-associated glycoproteins had a high degree of homology for different alleles of the two species but had no relationship to the N-terminal sequences of the S-allele-associated glycoproteins from N. alata[46] or L. peruvianum.[41] A feature of the Brassica S-allele-associated glycoproteins is the cluster of six disulphide bonds at the carboxy terminal end of the molecule which would give a rigid region;[48-50] there are also regions within the protein sequences which are relatively conserved and relatively variable. The most variable region was relatively hydrophilic. There was no homology with the sequence for the $\underline{S}_2$-glycoprotein of N. alata.[46] This raises the possibility that the sporophytic and gametophytic systems of self-incompatibility evolved separately.

Thus, we are accumulating precise information on the structures of the style S-allele-associated glycoproteins. This will eventually lead us to an understanding of the S-allele-associated molecules within the pollen and then to approaches for understanding function.

UNDERSTANDING THE MECHANISM OF SELF-INCOMPATIBILITY

We are starting to accumulate the molecular tools which will enable us to approach the questions of how different alleles are generated and how the corresponding gene products in pollen and pistil interact to cause arrest of incompatible pollen tubes. There are two main theories which address the mode of S-gene action (for reviews[4,7,9,53]). Either the growth of incompatible pollen tubes is inhibited (the oppositional hypothesis) or their growth is not stimulated (the complementary hypothesis). Mulcahy and Mulcahy[54] have proposed a radically different "heterosis" theory which discounts the existence of single locus, multiallelic self-incompatibility. It is built on the idea that several loci are involved and that pollen tubes grow rapidly if there are differences in many of these loci, and slowly if many of the alleles are the same in pollen and pistil. This theory has been challenged on a number of grounds.[55] No single theory accounts for all the observations made at the genetic and cytological level. Many contributors to this field have advanced complex biochemical bases for these theories. While these ideas are valuable in thinking and give a framework for experimental design, solid advances in understanding the subject will depend on understanding the structure and function of the products of different _S_-alleles in both pollen and style, as well as a general knowledge of the molecular species which are involved in nature and control of pollen tube growth.

One approach to understanding the function of the putative products of the _S_-gene in the style has been to study the effect of style extracts on the growth of pollen tube _in vitro_. In several cases specific inhibition of pollen tube growth by extracts of styles bearing the same allele as the test pollen has been reported.[40,56,57] There are difficulties in interpreting these experiments in terms of the above theories. One problem is that the _in vitro_-grown pollen is presented with a mixture of components; it is also hard to devise a medium for such experiments in the absence of information regarding essential co-factors and optimum conditions for both pollen tube growth and the action of isolated style components. The experimental difficulties are compounded by the requirement to measure the lengths of many pollen tubes, which is a time-consuming, tedious process. We have attempted to establish an assay for _in vitro_ pollen tube growth based on an ELISA test using a monoclonal antibody which binds to the α-L-arabinofuranosyl residues of the pollen tube wall.[58] We had expected that the amount of antibody bound would be directly proportional to the surface area and therefore the length of the pollen tube. However, we found that the relationship was logarithmic; that is, more antibody binds per unit length to short tubes than long tubes. Nonetheless, the assay was used to measure the effect of several chemical inhibitors of pollen tube growth.[58] When we attempted to apply this assay to measure the effect of the $\underline{S}_2$-allele-associated style glycoprotein we found that the level of antibody bound increased although the tube growth was inhibited by the $\underline{S}_2$-glycoprotein.[59] These experiments indicate that under the experimental conditions, a style component associated with a particular S-allele, isolated under mild conditions and which appears to be a single component, acts as an inhibitor of pollen tube growth in an allele-specific manner. Study of the action of this isolated $\underline{S}_2$-glycoprotein on pollen tube biosynthesis may give us new insights into the mechanism of self-incompatibility.

CONCLUSIONS

1. Self-incompatibility is a relatively simple model system for studying the mechanisms of recognition and response in plant cell interactions.

2. The application of recombinant DNA techniques to cloning S-allele-associated glycoproteins has given a major impetus to the field. Recombinant DNA studies will rapidly enhance our understanding of the generation of new alleles and of the nature of the S-gene products in the pollen.

3. Ultimately, our understanding of the genetic control and the biochemical expression of self-incompatibility will depend on having an integrated multidisciplinary approach to the problem. We rely on the classical techniques of plant breeding for the fundamental genetics of the systems. We draw heavily on the careful observations of many pollination biologists to design molecular approaches. Immunological and recombinant DNA techniques can be used to produce specific probes; the information we get from these probes is enhanced by using them in immunocytochemical and hybridization histochemistry studies to determine the cellular and subcellular expression of particular molecules. That is, the molecular and microscope studies must be interactive to be most effective.

4. Our data base on molecular aspects of self-incompatibility is very limited. We have studies of some aspects of a very few systems. In contrast, the

biology of many systems has been studied and the mechanisms for expression of self-incompatibility at the cellular level are apparently quite diverse. We should be cautious in drawing general conclusions from the study of a few systems.

REFERENCES

1. HARRIS, P.J., M.A. ANDERSON, A. BACIC, A.E. CLARKE. 1984. Cell-cell recognition in plants with special reference to the pollen-stigma interaction. In Oxford Surveys of Plant Molecular and Cell Biology. (B.J. Miflin, ed.), Vol. 1, Oxford University Press, pp. 161-203.
2. CLARKE, A.E., M.A. ANDERSON, T. BACIC, P.J. HARRIS, S-L. MAU. 1985. Molecular basis of cell recognition during fertilization in higher plants. J. Cell Sci. Suppl. 2: 261-285.
3. CLARKE, A.E., P.A. GLEESON. 1981. Molecular aspects of recognition and response in pollen-stigma interaction. Recent Adv. Phytochem. 15: 161-211.
4. de NETTANCOURT, D. 1977. Incompatibility in Angiosperms. Springer-Verlag, Berlin, 230 p.
5. LEWIS, D. 1949. Incompatibility in flowering plants. Biol. Rev. Camb. Philos. Soc. 24: 472-496.
6. HESLOP-HARRISON, J. 1975. Incompatibility and the pollen-stigma interaction. Annu. Rev. Plant Physiol. 26: 403-425.
7. GIBBS, P. 1986. Do homomorphic and heteromorphic self-incompatibility systems have the same sporophytic mechanisms? Plant Syst. Evol. 154: 285-323.
8. PANDEY, K.K. 1960. Evolution of gametophytic and sporophytic systems of self-incompatibility in angiosperms. Evolution 14: 98-115.
9. LAWRENCE, M.J. 1975. The genetics of self-incompatibility. Proc. R. Soc. Lond. B. Biol. Sci. 188: 275-285.
10. LINSKENS, H.F. 1975. Incompatibility in Petunia. Proc. R. Soc. Lond. B. Biol. Sci. 188: 299-311.
11. CAMPBELL, J.M., M.J. LAWRENCE. 1981. The population genetics of the self-incompatibility polymorphism in Papaver rhoeas. I. The number and distribution of S-alleles in families from three localities. Heredity 46: 69-79.

12. HESLOP-HARRISON, J. 1975. Male gametophyte selection and the pollen-stigma interaction. In Gametophyte Competition in Plants and Animals. (D. Mulcahy, ed.), Amsterdam: North Holland, pp. 177-190.
13. DICKINSON, H.G., J. LAWSON. 1975. Pollen tube growth in the stigma of Oenothera organensis following compatible and incompatible intraspecific pollinations. Proc. R. Soc. Lond. B. Biol. Sci. 188: 327-344.
14. SAVITSKY, H. 1950. A method for determining self-sterility in sugar beets based upon the stage of ovule development shortly after flowering. Proc. Am. Soc. Sugar Beet Tech. 1950: 198-201.
15. BREDEMEIJER, G.M.M., J. BLAAS. 1981. S-Specific proteins in styles of self-incompatible Nicotiana alata. Theor. Appl. Genet. 59: 185-190.
16. PANDEY, K.K. 1967. Elements of the S-gene complex. II. Mutation and complementation at the S-I locus in Nicotiana alata. Heredity 22: 255-283.
17. EVANS, N.A., B.A. STONE. 1984. Characteristics and specificity of the interaction of a fluorochrome from aniline blue (Sirofluor) with polysaccharides. Carbohydr. Polym. 4: 215-230.
18. STONE, B.A., N.A. EVANS, I. BONIG, A.E. CLARKE. 1984. The application of Sirofluor, a chemically defined fluorochrome from aniline blue, for the histochemical detection of callose. Protoplasma 122: 191-195.
19. RAE, A.L., P.J. HARRIS, A. BACIC, A.E. CLARKE. 1985. Composition of the cell walls of Nicotiana alata Link et Otto pollen tubes. Planta 166: 128-133.
20. DARVILL, A.G., M. McNEIL, P. ALBERSHEIM, D.P. DELMER. 1980. The primary cell walls of flowering plants. In The Biochemistry of Plants. (N.E. Tolbert, ed.), Vol. 1, Academic Press, London, pp. 92-162.
21. ANDERSON, M.A., M.S. SANDRIN, A.E. CLARKE. 1984. A high proportion of hybridomas raised to a plant extract secrete antibody to arabinose or galactose. Plant Physiol. 75: 1013-1016.
22. ANDERSON, M.A., P.J. HARRIS, I. BONIG, A.E. CLARKE. 1987. Immuno-gold localization of α-L-arabinofuranosyl residues in pollen tubes of Nicotiana alata Link et Otto. Planta 171: 438-442.
23. DELMER, D.P. 1987. Cellulose biosynthesis. Annu. Rev. Plant Physiol. 38: 259-290.

24. GLEESON, P.A., A.E. CLARKE. 1979. Structural studies on the major component of the Gladiolus style mucilage, an arabinogalactan-protein. Biochem. J. 181: 607-621.
25. LABARCA, C., F. LOEWUS. 1972. The nutritional role of pistil exudate in pollen tube wall formation in Lilium longiflorum. I. Utilization of injected stigmatic exudate. Plant Physiol. 50: 7-14.
26. ASPINALL, G.O., K.G. ROSELL. 1978. Polysaccharide component in the stigmatic exudate from Lilium longiflorum. Phytochemistry 17: 919-921.
27. HOGGART, R.M., A.E. CLARKE. 1986. Arabinogalactans are common components of Angiosperm styles. Phytochemistry 23: 1571-1573.
28. CLARKE, A.E., R. ANDERSON, B.A. STONE. 1979. Form and function of arabinogalactans and arabino-galactan proteins. Phytochemistry 18: 521-540.
29. FINCHER, G.B., B.A. STONE, A.E. CLARKE. 1983. Arabinogalactan-proteins: structure, biosynthesis and function. Annu. Rev. Plant Physiol. 34: 47-70.
30. SEDGLEY, M., M.A. BLESING, I. BONIG, M.A. ANDERSON, A.E. CLARKE. 1986. Arabinogalactan-proteins are localized extracellularly in the transmitting tissue of Nicotiana alata Link and Otto, an ornamental tobacco. Micron. Microsc. Acta 16: 247-254.
31. VAN HOLST, G-J., A.E. CLARKE. 1985. Quantification of arabinogalactan-protein in plant extracts by single radial gel diffusion. Anal. Biochem. 148: 446-450.
32. VAN HOLST, G-J., A.E. CLARKE. 1986. Organ-specific arabinogalactan-proteins of Lycopersicon peruvianum (Mill) demonstrated by crossed-electrophoresis. Plant Physiol. 80: 786-798.
33. GELL, A.C., A. BACIC, A.E. CLARKE. 1986. Arabino-galactan-proteins of the female sexual tissue of Nicotiana alata. I. Changes during flower development and pollination. Plant Physiol. 82: 885-887.
34. BACIC, A., A.C. GELL, A.E. CLARKE. 1987. Structural studies on a major component of Nicotiana alata stigmas: arabinogalactan-proteins. Phytochemistry 26: (in press).
35. LEWIS, D. 1952. Serological reactions of pollen incompatibility substances. Proc. R. Soc. Lond.

B. Biol. Sci. 140: 127-135.
36. RAFF, J.W., R.B. KNOX, A.E. CLARKE. 1981. Style antigens of Prunus avium. Planta 153: 125-129.
37. MAU, S-L., J. RAFF, A.E. CLARKE. 1982. Isolation and partial characterization of components of Prunus avium L. styles, including an antigenic glycoprotein associated with a self-incompatibility genotype. Planta 156: 505-516.
38. WILLIAMS, E.G., S. RAMM-ANDERSON, C. DUMAS, S-L. MAU, A.E. CLARKE. 1982. The effect of isolated components of Prunus avium L. styles on in vitro growth of pollen tubes. Planta 156: 517-519.
39. KAMBOJ, R.K., J.F. JACKSON. 1986. Self-incompatibility alleles control a low molecular weight, basic protein in pistils of Petunia hybrida. Theor. Appl. Genet. 71: 815-819.
40. DICKINSON, H.G., J. MORIARTY, J. LAWSON. 1982. Pollen-pistil interaction in Lilium longiflorum: the role of the pistil in controlling pollen tube growth following cross pollination and self-pollination. Proc. R. Soc. Lond. B. Biol. Sci. 215: 45-62.
41. MAU, S-L., E.G. WILLIAMS, A. ATKINSON, M.A. ANDERSON, E.C. CORNISH, B. GREGO, R.J. SIMPSON, A. KHEYR-POUR, A.E. CLARKE. 1986. Style proteins of a wild tomato associated with expression of self-incompatibility. Planta 169: 184-191.
42. NASRALLAH, M.E., J.T. BARBER, D.H. WALLACE. 1970. Self-incompatibility proteins in plants: detection, genetics and possible mode of action. Heredity 25: 25-27.
43. NASRALLAH, J.B., M.E. NASRALLAH. 1984. Electrophoretic heterogeneity exhibited by S-allele specific glycoproteins of Brassica. Experentia 40: 279-281.
44. NISHIO, T., K. HINATA. 1977. Analysis of S-specific proteins in stigmas of Brassica oleracea L. by isoelectric focussing. Heredity 38: 391-396.
45. NISHIO, T., K. HINATA. 1982. Comparative studies on S-glycoproteins purified from different S-genotypes in self-incompatible Brassica species. I. Purification and chemical properties. Genetics 100: 641-647.
46. ANDERSON, M.A., E.C. CORNISH, S-L. MAU, E.G. WILLIAMS, R. HOGGART, A. ATKINSON, I. BONIG, B. GREGO, R. SIMPSON, P.J. ROCHE, J.D. HALEY, J.D. PENSCHOW,

H.D. NIALL, G.W. TREGEAR, J.P. COGHLAN, R.J. CRAWFORD, A.E. CLARKE. 1986. Cloning of cDNA for a stylar glycoprotein associated with expression of self-incompatibility in Nicotiana alata. Nature 321: 38-44.

47. NASRALLAH, J.B., T-H. KAO, M.L. GOLDBERG, M.E. NASRALLAH. 1985. A cDNA clone encoding an S-locus-specific glycoprotein from Brassica oleracea. Nature 318: 263-267.

48. NASRALLAH, J.B., T-H. KAO, C-H. CHEN, M.L. GOLDBERG, M.E. NASRALLAH. 1987. Amino-acid sequence of glycoproteins encoded by three alleles of the S locus of Brassica oleracea. Nature 326: 617-619.

49. TAKAYAMA, S., A. ISOGAI, C. TSUKAMOTO, Y. UEDA, H. HINATA, K. OKAZAKI, A. SUZUKI. 1986. Isolation and some characterization of S-locus-specific glycoproteins associated with self-incompatibility in Brassica campestris. Agric. Biol. Chem. 50(5): 1365-1367.

50. TAKAYAMA, S., A. ISOGAI, C. TSUKAMOTO, Y. UEDA, K. HINATA, I. OKAZAKI, A. SUZUKI. 1987. Sequences of S-glycoproteins, products of the Brassica campestris self-incompatibility locus. Nature 326: 102-105.

51. CORNISH, E.C., J.M. PETTITT, I. BONIG, A.E. CLARKE. 1987. Developmentally-controlled, tissue-specific expression of a gene associated with self-incompatibility in Nicotiana alata. Nature 326: 99-102.

52. CLARKE, A.E., M.A. ANDERSON, A. BACIC, E.C. CORNISH, P.J. HARRIS, S-L. MAU, J.R. WOODWARD. 1987. Molecular aspects of self-incompatibility in flowering plants. UCLA Symposia of Molecular and Cellular Biology: Plant Gene Systems and their biology. Alan R. Liss, Inc., New York (in press).

53. LEWIS, D. 1960. Genetic control of specificity and activity of the S antigen in plants. Proc. R. Soc. Lond. B. Biol. Sci. 151: 468-477.

54. MULCAHY, D.L. G.B. MULCAHY. 1983. Gametophytic self-incompatibility re-examined. Science 220: 1247-1251.

55. LAWRENCE, M.J., D.F. MARSHALL, V.E. CURTIS, C.H. FEARON. 1985. Gametophytic self-incompatibility re-examined: a reply. Heredity 54: 131-138.

56. BREWBAKER, J.L., S.K. MAJUMDER. 1961. Cultural studies of the pollen population effect and the

self-incompatibility inhibition. Am. J. Bot. 48: 457-464.

57. TOMKOVA, J. 1959. Problems of autosterility in _Nicotiana alata_, Link et Otto (formation of inhibitory factor in the pistil). Biol. Plant 1: 328-329.

58. HARRIS, P.J., K. FREED, M.A. ANDERSON, J.A. WEINHANDL, A.E. CLARKE. 1987. An enzyme-linked immunosorbent assay (ELISA) for _in vitro_ pollen growth based on binding of a monoclonal antibody to the pollen tube surface. Plant Physiol. 84: 851-855.

59. HARRIS, P.J., J.A. WEINHANDL, A.E. CLARKE. 1987. The effects of an isolated style glycoprotein from _Nicotiana alata_ on the _in vitro_ growth of pollen tubes (in preparation).

Chapter Five

THE ROLE OF PLANT COMPOUNDS IN THE REGULATION OF RHIZOBIUM NODULATION GENES

N. KENT PETERS AND SHARON R. LONG

Department of Biological Sciences
Stanford University
Stanford, California 94305

INTRODUCTION

Members of the genus Rhizobium are soil bacteria capable of forming symbiotic nitrogen fixing nodules with leguminous plants. Individual species of bacteria have a defined host-range; they are able to form nodules on a limited number of plant species.[1] The bacteria invade the plant root through an emerging root hair causing the root hair to curl. The bacteria penetrate through several cell layers into the root via an infection thread which is a tube-like structure of plant cell wall.[2] Early during the invasion process, in advance of the invading infection thread, inner cortical cells are stimulated to divide by the bacteria.[3] It is apparent from this process that there is an ordered and specific exchange of biochemical signals between the plant and the invading bacteria.

A few plant genes and enzyme functions involved in the symbiosis are known, such as plant leghemoglobin, uricase and glutamine synthetase, and bacterial nitrogenase.[4] Much work has focused on the bacterial genetics of this symbiosis and ten or more bacterial genes required for nodule formation have been identified by transposon mutagenesis.[5,6] Several of these genes have been sequenced[6,7] and the translation products identified,[8,9,10] but no function has been assigned to any of the genes. A few of these genes (nodABC) are able to complement mutants in other bacterial species and are known as common nodulation genes.[11,12,13] Other genes have been identified which affect the efficiency of nodulation or change the host range of the Rhizobium.[14,15]

An early suggestion that the bacteria could be conditioned or influenced by the plant to initiate nodules came from Bhagwat and Thomas who demonstrated that a delay in nodule initiation, characteristic of one strain of Rhizobium, could be elimited by culturing the bacteria in root exudate prior to inoculation.[16] Using a nodABC-lacZ translation fusion, Mulligan and Long[17] were first to show that the common nod genes, nodABC, of Rhizobium meliloti were inducible by host and non-host plant exudate and that the nodD gene product was required for induction. It has since been shown that the common nod genes of Rhizobium trifolii[15] and Rhizobium leguminosarum[18] are also regulated in this way. In addition, the R. trifolii[15] and R. leguminosarum[19] nodulation genes involved in determining host range were found to be inducible. A common feature of the genes induced by plant factors was shown to be a conserved region of 25 base pairs 200 to 240 base pairs upstream from the translation start site.[20] Given the involvement of the plant factor in regulating the expression of the bacterial nodulation genes, it was of obvious importance to isolate and identify the compound or compounds exuded by the plant root to which the bacteria respond.[21]

ISOLATION AND IDENTIFICATION OF HOST PLANT FACTOR

Isolation of the compound(s) in plant exudate which induce the nodulation genes was facilitated by an easy and rapid assay for induction using the nodABC-lacZ translational fusion.[17] Alfalfa seed exudate was fractionated by

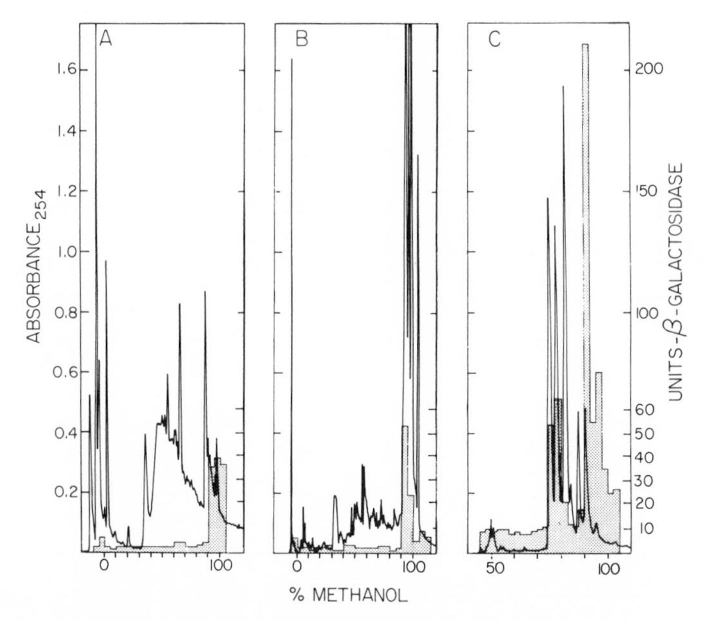

Fig. 1. Fractionation of exudate and ether extracted compounds by reverse-phase liquid chromatography. The elution profile as monitored by absorption at 254 nm is shown as a solid line for aqueous exudate (A) and ether extracted compounds (B and C). The percent of methanol in the elution gradient is given on the horizontal axis. The β-galactosidase activity assayed in each fraction is shown as a shaded bar graph.

HPLC and assayed for inducing activity (Fig. 1A). The inducing activity eluted from the column between 95 to 100% methanol demonstrating its hydrophobic character. The inducing activity could be partially purified by partitioning the inducing compounds into diethyl ether (Fig. 1B). The ether soluble materials could be separated and the majority of the inducing activity co-eluted with a single UV absorbing peak between 90 to 95% methanol (Fig. 1C). Fractions containing the inducing activity were repurified and pooled for structural determination.

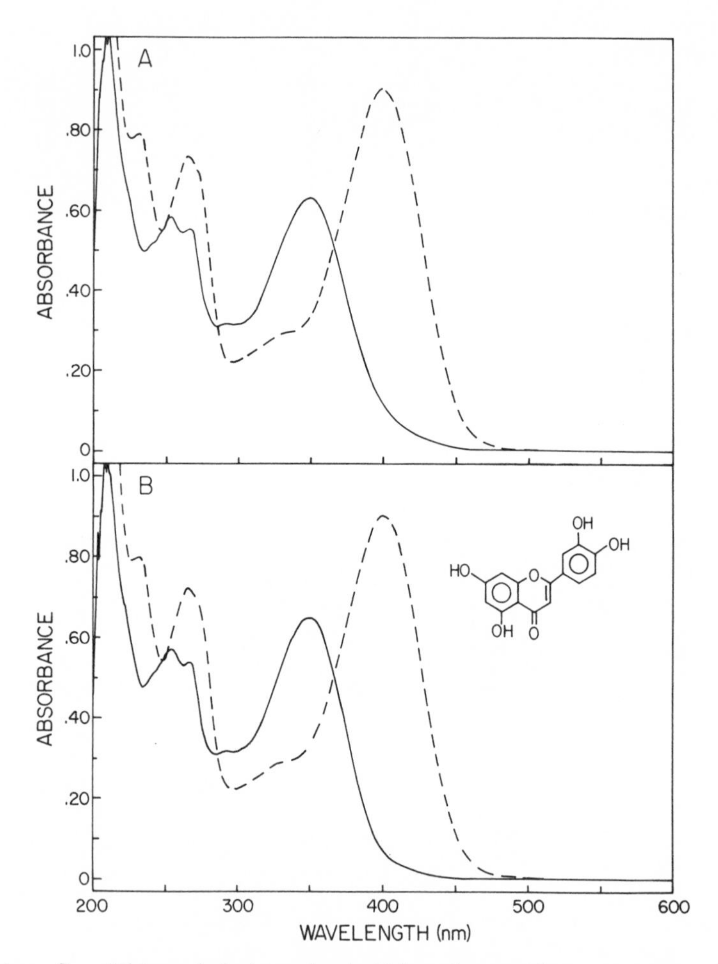

Fig. 2. Ultraviolet and visible absorption spectra of purified inducer and synthetic luteolin. The absorption spectra of purified inducer (A) and synthetic luteolin (B) are depicted as measured in methanol (——) and in methanol/NaOH (- -). The structure of luteolin is shown in B.

Table 1. Absorbance maxima in nm.

Solvent	Band I		Band II	
	Luteolin	Inducer	Luteolin	Inducer
MeOH	349	351	253,267	256,267
MeOH/NaOH	401	405	266	266
MeOH/$AlCl_3$	328,426	328,426	271	274
MeOH/$AlCl_3$/HCl	355	357	274	275
MeOH/NaAc	384	384	269	269
MeOH/NaAc/H_3BO_3	370	372	259	259

Spectral analyses were performed as described by Mabry *et al.*[23]

The key to determining the chemical identity of the inducer was its absorption spectrum in ultraviolet and visible light. The inducer showed two major absorbance maxima at 256 and 351 nm (Fig. 2 and Table 1). These absorbance maxima were found to change in the presence of strong base (Fig. 2 and Table 1). These spectra were compared to the spectra of various plant secondary metabolites and found to be typical of flavonoid molecules. By determining the absorption spectrum of the inducer in a series of solvents (Table 1) and using the guidelines for determination of flavonoid structure described by Mabry, Markham and Thomas[22] the compound was determined to be the plant flavone, 3',4',5,7-tetrahydroxyflavone known as luteolin.

The isolated inducer was further analyzed by mass spectrometry and proton nuclear magnetic resonance spectroscopy. The mass spectrum shows a molecular ion M^+ (m/e = 286) and a base peak (m/e = 153) (Fig. 3). From the known fragmentation of flavonoids,[23] these are the expected molecular ion and base peak for luteolin. The nuclear magnetic resonance spectrum indicated the presence

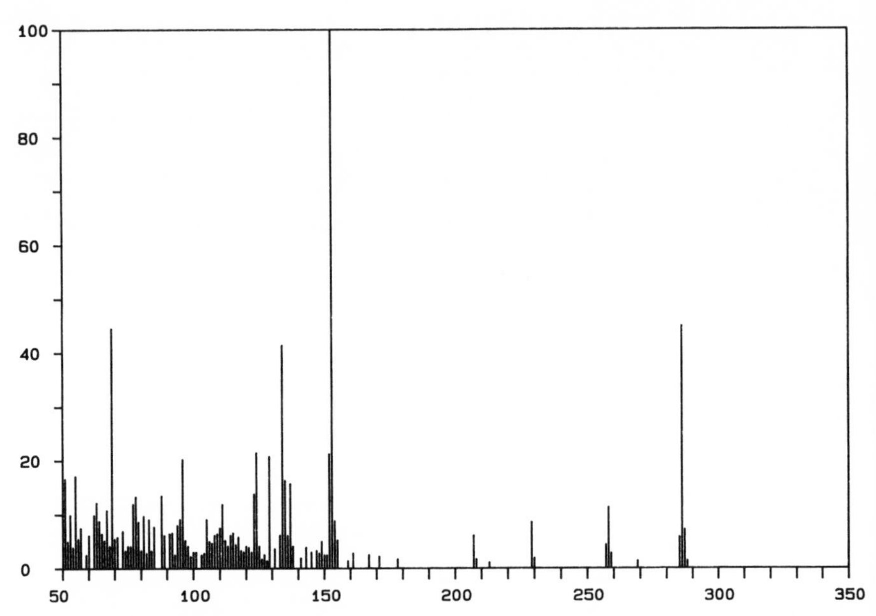

Fig. 3. Mass spectrum of isolated inducer. The frequency of an ion is plotted against the ion's mass/charge ratio.

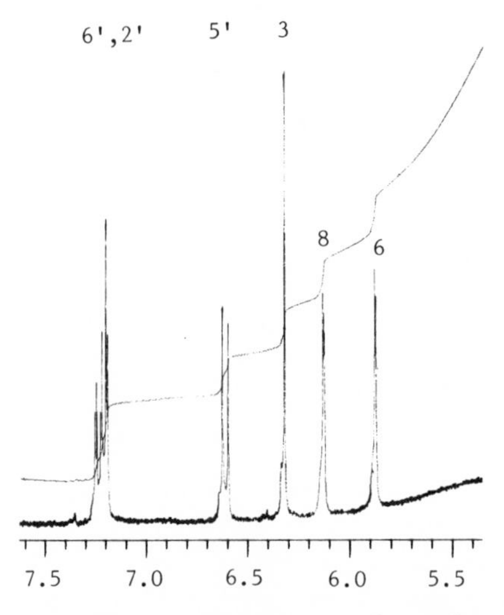

Fig. 4. Proton magnetic resonance of isolated inducer. Assignments of the proton positions are given above the resonances.

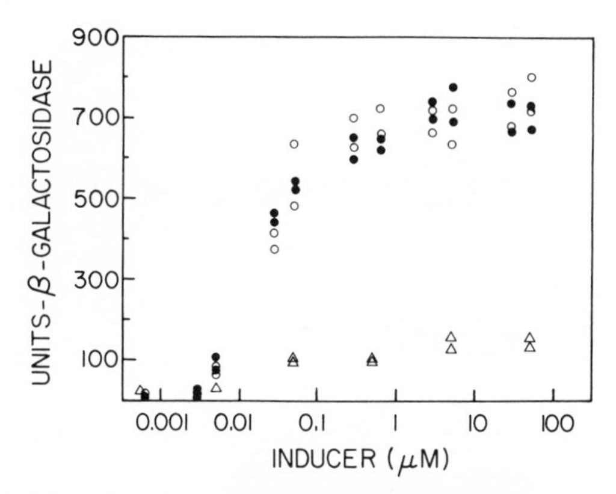

Fig. 5. Comparison of inducing activity of purified inducer and synthetic luteolin. The relative units of β-galactosidase activity are plotted on the vertical axis versus the concentration of purified inducer (O), synthetic luteolin (●) and apigenin (Δ).

of six aromatic or vinyl protons, also consistent with its identification as luteolin (Fig. 4).

Since the isolated inducer has the physical characteristics of luteolin, authentic luteolin was assayed for its inducing activity. Figure 5 compares the activity of authentic luteolin and isolated inducer over a range of concentrations from 0.001 to 100 μM. The concentration dependence of nodABC induction on isolated inducer and synthetic luteolin is the same. This confirmed that luteolin is the active inducer present in plant exudate.

Two points about the concentration dependence of nodABC induction should be noted. First, the induction profile is similar to those of the inducible lactose[24] and arabinose[25] operons of *Escherichia coli* in that the induction curve is sigmoidal on a linear/logarithmetric plot of inducer concentration. Second, the concentration of luteolin required to give half maximal induction for the nodABC operon is three to five orders of magnitude less than the concentration of inducer required for the lactose or arabinose operons. This greater sensitivity may reflect the availability of inducer in the rhizosphere and is

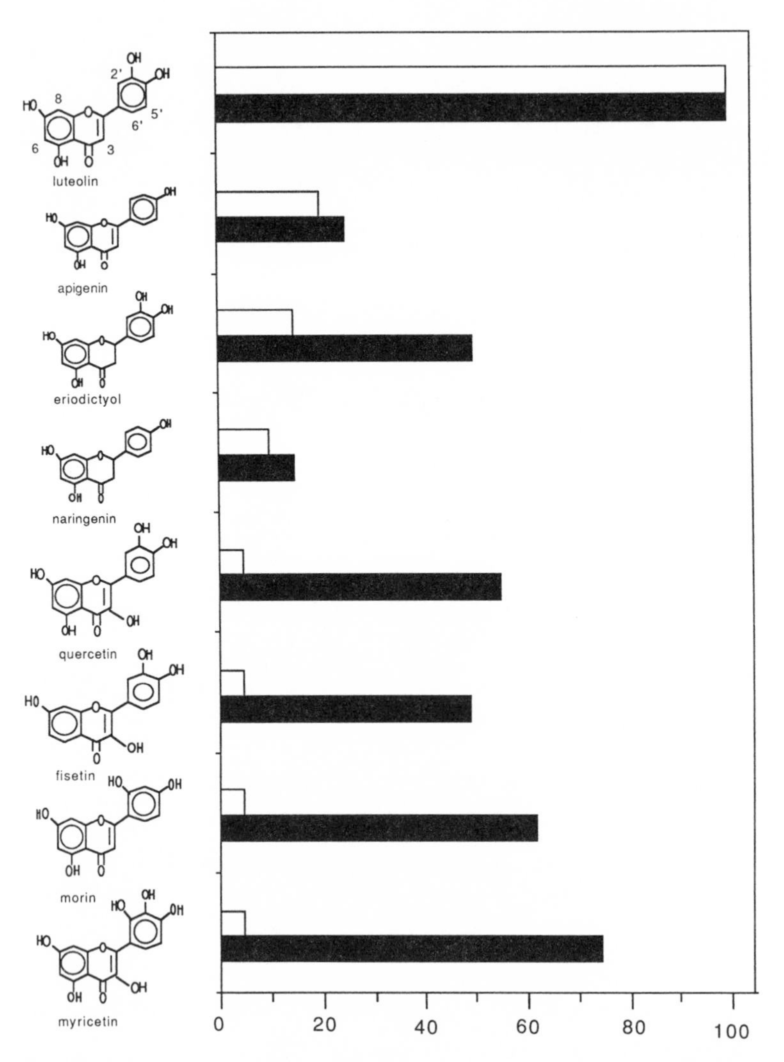

Fig. 6. Inducing and antagonizing activity of various flavonoids. A compound's inducing activity at 1 µM is given as percent activity of luteolin at 1 µM (open bar). The percent of full induction by 1 µM luteolin in the presence of 1 µM of the indicated compound (solid bar).

indicative of the operon's function. The lactose and arabinose operons are operons which are expressed to utilize carbon and energy sources. These operons are only expressed in the presence of significant concentrations of substrate. The nodABC operon by contrast is responding to a developmental signal which need not be present at a high concentration.

ACTIVITY OF STRUCTURALLY RELATED COMPOUNDS

Luteolin is just one member of a wide variety of flavonoid molecules. It thus was interesting to test the activities of some structurally related compounds. Compounds structurally related to luteolin were purchased and assayed for inducing activity at 1 μM. Of the several compounds tested (flavone, flavonol, chrysin, fisetin, morin, myricetin, quercetin, apigenin, eriodictyol, dihydroquercetin and naringenin), only apigenin, naringenin and eriodictyol gave detectable inducing activity (Fig. 6). These results demonstrate certain structural requirements for the inducer of the *R. meliloti* nodulation genes. To have activity, the molecule must have a 4' hydroxy substituent. A hydroxyl at the 3 or 2' positions (quercetin and morin, respectively) can abolish activity, and a saturated bond between C2 and C3 (eriodictyol), which creates a chiral center at C2 and puckers the molecule, greatly decreases activity. The structural requirements for a flavonoid to work as an inducer of the *R. meliloti* nodABC are quite specific and any variation from the luteolin molecule reduces the activity.

FLAVONOIDS AS INDUCERS FOR OTHER *RHIZOBIUM* SPECIES

Inducing compounds from other host-*Rhizobium* systems have been isolated or otherwise established. In each of these cases, the inducing molecules have been identified as flavonoids. Redmond *et al.*[26] isolated 4',7-dihydroxyflavone from clover root exudates as that compound which best induced the nodulation genes of its symbiont, *R. trifolii* (Fig. 7). In *R. trifolii*, the dihydroxyflavone and apigenin both induce better than luteolin. Thus a 3' hydroxy reduces the activity as an inducer for *R. trifolii*, but is required for full activity with *R. meliloti*. Firmin *et al.*[27] tested several flavonoid molecules and found that

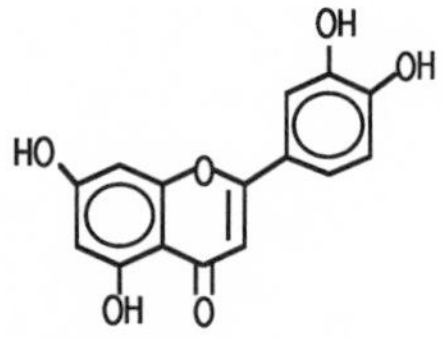

LUTEOLIN

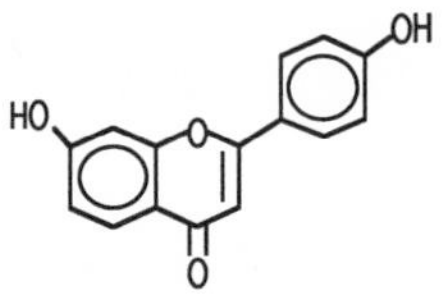

DIHYDROXYFLAVONE 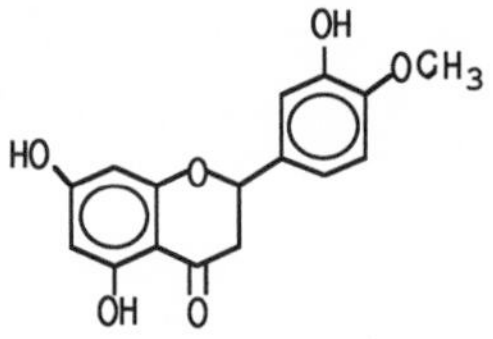 HESPERITIN

Fig. 7. Structures of *Rhizobium* nodulation gene inducers.

3',5,7-trihydroxy-4'-methoxyflavanone (hesperitin) and 3',4',5,7-tetrahydroxyflavanone (eriodictyol) best induced the nodulation genes of *R. leguminosarum* (Fig. 6 and 7). It is interesting that flavanones induce *R. leguminosarum* best since flavanones induce poorly for both *R. meliloti* and *R. trifolii*.

ANTAGONISTIC ACTION OF STRUCTURALLY RELATED COMPOUNDS

Molecules which differ from luteolin by a single hydroxyl substitution or double bond have considerably reduced activity. We tested these molecules for their ability to antagonize the induction of the *nodABC* genes by luteolin. Induction by luteolin at a concentration capable of giving near full induction (1 μM) was assayed in the presence of 1 μM of a structurally related compound (Fig. 6). All of the compounds tested were found to antagonize the induction by luteolin. The best of these antagonists was naringenin. This molecule lacks the 3' hydroxyl and has a saturated C2-C3 bond. These two features make this molecule unable to induce significantly, but it is an effective antagonist of induction at concentrations equimolar to luteolin. The most reasonable interpretation of this antagonistic property is that the naringenin (and other flavonoids) competes with luteolin for a common binding site.

Firmin *et al.*[27] have also shown that flavonoid and other molecules which can not induce the nodulation genes can antagonize the activation by suitable inducer. Compounds which had substitutions at the 3 position such as kaempferol and rhamnetin were inhibitory. This is similar to our observations and those of Djordjevic *et al.*[28] Interestingly, Firmin *et al.*[27] also found that compounds that induce the *Agrobacterium tumefaciens vir*[27] genes can inhibit induction of the nodulation genes. Other compounds which have been shown to inhibit activity are some isoflavones and 4-hydroxycoumarin (umbelliferone) which were found in root exudates of clover.[28] Djordjevic *et al.*[28] have shown that some regions of a growing root exude inhibitory compounds, but the physiological role or effect on nodulation has not been characterized.

ROLE OF INDUCER AND *nodD* GENE PRODUCT IN HOST RANGE

Among the species *R. meliloti*, *R. trifolii* and *R. leguminosarum* the inducer and its interaction with the *nodD* gene product does not determine host range. This is evidenced by the fact that the *R. meliloti nodD* gene can complement a *R. trifolii nodD* mutant with no change in host range.[11] *R. meliloti* which cannot nodulate *Macroptilium atropurpureum* (siratro). However it can nodulate siratro when it carries a plasmid with the *nodD* from the broad host range *Rhizobium* NGR234, which can nodulate siratro.[30,31] The *nodD* gene may not be the only gene required, however, since the DNA fragment used also contains another genetic locus. This other locus is required for the extension of *R. trifolii* host range to siratro.[32] The *nodD* from NGR234 was not needed to extend the host range of *R. trifolii* to siratro.[32] The *nodD* gene can affect response to plant exudate and may therefore affect host range, but *nodD* has not been demonstrated to be the only factor involved.

ROLE OF INDUCER IN NODULATION EFFICIENCY

Evidence that the level of inducer made by the plant is important to nodulation efficiency comes from a study by Kapulnik et al.[33] These workers have found that a line of alfalfa selected for greater nitrogen fixation and growth has sixty percent greater inducing activity than the unselected line. In addition, Kapulnik *et al*. found that watering alfalfa with 10 μM luteolin could double the number of nodules formed per plant. This suggests that the level of induction of the *Rhizobium* by the host plant can be rate limiting.

CONCLUSIONS

Rhizobium nodulation genes are induced by plant flavonoids. The flavonoids isolated from different plants have different structures, but are either flavones or flavanones. Among *R. meliloti*, *R. trifolii* and *R. leguminosarum*, the inducer specificity does not affect host range, but inducer specificity may be partially responsible for affecting host range. Molecules structurally related to inducing molecules have been shown to antagonize induction by known inducers, but the physiological significance of this is not understood. The amount of inducer produced by the host plant can affect the efficiency of nodulation. Therefore the amount and the structure of the flavonoids produced by the plant play a critical role in the establishment of the symbiosis.

ACKNOWLEDGMENTS

This work was supported by NIH grant R01-GM30962 and Department of Energy contract AS03-83-ER12084 to S.R.L. N.K.P. was supported by an NSF Plant Biology Postdoctoral Fellowship. We are grateful to J. Frost and L. Reimer for their help with the isolation and chemical identification of the inducer.

REFERENCES

1. VINCENT, J.M. 1977. *Rhizobium*: general microbiology. *In* A Treatise on Dinitrogen Fixation, Sec. III: Biology. (R.W.F. Hardy, W.S. Silver, eds.), Wiley, New York, pp. 277-366.
2. LONG, S.R. 1984. Genetics of *Rhizobium* nodulation. *In* Plant-Microbe Interactions. (T. Kosuge, E. Nester, eds.), Macmillan, New York, pp. 256-306.

3. DUDLEY, M.E., T.W. JACOBS, S.R. LONG. 1987. Microscopic studies of cell division induced in alfalfa roots by Rhizobium meliloti. Planta, in press.
4. VERMA, D.P.S., K. NADLER. 1984. Legume-Rhizobium-symbiosis: host's point of view. In Plant Gene Research. (D.P.S. Verma, T.H. Hohn, eds.), Springer-Verlag, New York, pp. 58-93.
5. LONG, S.R., W.J. BUIKEMA, F.M. AUSUBEL. 1982. Cloning of Rhizobium meliloti nodulation genes by direct complementation of nod$^-$ mutants. Nature 298: 485-488.
6. JACOBS, T.W., T.T. EGELHOFF, S.R. LONG. 1985. Physical and genetic map of Rhizobium meliloti nodulation gene region and nucleotide sequence of nodC. J. Bacteriol. 162: 469-476.
7. EGELHOFF, T.T., R.F. FISHER, T.W. JACOBS, J.T. MULLIGAN, S.R. LONG. 1985. Nucleotide sequence of Rhizobium meliloti 1021 nodulation genes: nodD is read divergently from nodABC. DNA 4: 241-249.
8. EGELHOFF, T.T., S.R. LONG. 1985. Rhizobium meliloti nodulation genes: identification of nodABC gene products, purification of nodA protein, and expression of nodA in Rhizobium meliloti. J. Bacteriol. 164: 591-599.
9. SCHMIDT, J., M. JOHN, E. KONDOROSI, A. KONDOROSI, U. WIENEKE, G. SCHRODER, J. SCHRODER, J. SCHELL. 1984. Mapping of the protein-coding regions of Rhizobium meliloti common nodulation genes. EMBO J. 3: 1705-1711.
10. DOWNIE, J.A., C.D. KNIGHT, A.W.B. JOHNSTON, L. ROSSEN. 1985. Identification of genes and gene products involved in the nodulation of peas by Rhizobium leguminosarum. Mol. Gen. Genet. 198: 255-262.
11. FISHER, R.F., J.K. TU, S.R. LONG. 1985. Conserved nodulation genes in Rhizobium meliloti and Rhizobium trifolii. Appl. Environ. Microbiol. 49: 1432-1435.
12. DJORDJEVIC, M.A., W. ZURKOWSKI, J. SHINE, B.G. ROLFE. 1983. Sym plasmid transfer to various symbiotic mutants of Rhizobium trifolii, Rhizobium leguminosarum, and Rhizobium meliloti. J. Bacteriol. 156: 1035-1045.
13. KONDOROSI, E., Z. BANFALVI, A. KONDOROSI. 1984. Physical and genetic analysis of a symbiotic region of Rhizobium meliloti: identification of nodulation genes. Mol. Gen. Genet. 193: 445-452.

14. PUTNOKY, P., A. KONDOROSI. 1986. Two gene clusters of Rhizobium meliloti code for early essential nodulation functions and a third influence nodulation efficiency. J. Bacteriol. 167: 881-887.
15. INNES, R.W., P.L. KUEMPEL, J. PLAZINSKI, H. CANTER-CREMERS, B.G. ROLFE, M.A. DJORDJEVIC. 1985. Plant factors induce expression of nodulation and host-range genes in Rhizobium trifolii. Mol. Gen. Genet. 201: 426-432.
16. BHAGWAT, A.A., J. THOMAS. 1982. Legume-Rhizobium interactions: cowpea root exudate elicits faster nodulation response by Rhizobium species. Appl. Environ. Microbiol. 43: 800-805.
17. MULLIGAN, J.T., S.R. LONG. 1985. Induction of Rhizobium meliloti nodC expression by plant exudate requires nodD. Proc. Natl. Acad. Sci. USA 82: 6609-6613.
18. ROSSEN, L., C.A. SHEARMAN, A.W.B. JOHNSTON, J.A. DOWNIE. 1985. The nodD gene of Rhizobium leguminosarum is autoregulatory and in the presence of plant exudates induces the nodA,B,C genes. EMBO J. 4: 3369-3373.
19. SHEARMAN, C.A., L. ROSSEN, A.W.B. JOHNSTON, J.A. DOWNIE. 1986. The Rhizobium leguminosarum nodulation gene nodF encodes a polypeptide similar to an acyl-carrier protein and is regulated by nodD plus a factor in pea root exudate. EMBO J. 5: 647-652.
20. ROSTA, K., E. KONDOROSI, B. HORVATH, A. SIMONCSITS, A. KONDOROSI. 1986. Conservation of extended promoter regions of nodulation genes in Rhizobium. Proc. Natl. Acad. Sci. USA 83: 1757-1761.
21. PETERS, N.K., J.W. FROST, S.R. LONG. 1986. A plant flavone, luteolin, induces expression of Rhizobium meliloti nodulation genes. Science 233: 977-980.
22. MABRY, T.J., K.R. MARKHAM, M.B. THOMAS. 1970. The structural analysis of flavonoids by ultraviolet spectroscopy. In The Systematic Identification of Flavonoids. Springer-Verlag, New York, pp. 35-175.
23. MABRY, T.J., K.R. MARKHAM. 1975. Mass spectroscopy of flavonoids. In The Flavonoids. (J.B. Harborne, T.J. Mabry, H. Mabry, eds.), Vol. 1, Academic Press, New York, pp. 78-126.
24. NOVICK, A., J.R. SADLER. 1965. The properties of repressor and the kinetics of its action.

J. Mol. Biol. 12: 305-327.
25. DOYLE, E.M., C. BROWN, R.W. HOGG, R.B. HELLING. 1972. Induction of the ara operon of Escherichia coli B/r. J. Bacteriol. 110: 56-65.
26. REDMOND, J.W., M. BATLEY, M.A. DJORDJEVIC, R.W. INNES, P.L. KUEMPEL, B.G. ROLFE. 1986. Flavones induce expression of nodulation genes in Rhizobium. Nature 323: 632-635.
27. FIRMIN, J.L., K.E. WILSON, L. ROSSEN, A.W.B. JOHNSTON. 1986. Flavonoid activation of nodulation genes in Rhizobium reversed by other compounds present in plants. Nature 324: 90-92.
28. DJORDJEVIC, M.A., J.W. REDMOND, M. BATLEY, B.G. ROLFE. 1987. Clover secrete specific phenolic compounds which either stimulate or repress nod gene expression in Rhizobium trifolii. EMBO J. 6: 1173-1179.
29. STACHEL, S.E., E. MESSENS, M. VAN MONTAGU, P. ZAMBRYSKI. 1985. Identification of the signal molecules produced by wounded plant cells that activate T-DNA transfer in Agrobacterium tumefaciens. Nature 318: 624-629.
30. BACHEM, C.W.B., Z. BANFALVI, E. KONDOROSI, J. SCHELL, A. KONDOROSI. 1986. Identification of host range determinants in the Rhizobium species MPIK3030. Mol. Gen. Genet. 203: 42-48.
31. HORVATH, B., C.W.B. BACHEM, J. SCHELL, A. KONDOROSI. 1987. Host-specific regulation of nodulation genes in Rhizobium is mediated by a plant-signal, interacting with the nodD gene product. EMBO J. 6: 841-848.
32. BASSAM, B.J., B.G. ROLFE, M.A. DJORDJEVIC. 1986. Macroptilium atropurpureum (siratro) host specific genes are linked to a nodD like gene in the broad host range Rhizobium strain NGR234. Mol. Gen. Genet. 203: 49-57.
33. KAPULNIK, Y., C.M. JOSEPH, D.A. PHILLIPS. 1987. Flavone limitations to root nodulation and symbiotic nitrogen fixation in alfalfa. Plant Physiol. 84: 1193-1196.

Chapter Six

INITIAL INTERACTIONS BETWEEN PLANT CELLS AND AGROBACTERIUM TUMEFACIENS IN CROWN GALL TUMOR FORMATION

GERARD A. CANGELOSI AND EUGENE W. NESTER

Department of Microbiology
University of Washington
Seattle, Washington 98195

INTRODUCTION

Crown gall tumors on dicotyledonous plants are formed by infection of wounded tissue with the bacterial plant pathogen Agrobacterium tumefaciens. Infection of the plant cells does not occur by the bacterial cells themselves, but by a discreet portion of a bacterial, tumor-inducing (Ti) plasmid (Fig. 1). This segment of DNA, termed T-DNA, is found integrated into the nuclear DNA of cloned plant tumor tissue.[1,2,3] Expression of T-DNA genes alters the hormone balance of the plant cells,[4,5,6] resulting in the tumor phenotype. T-DNA gene products also direct the transformed plant cells to produce opines, unusual amino acids which are utilized by the Agrobacteria.[7]

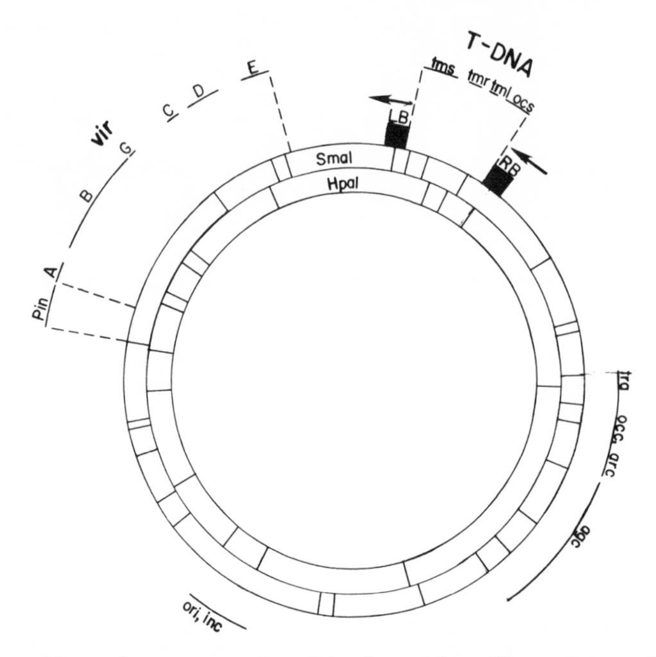

Fig. 1. Map of an octopine Ti plasmid. The vir region is in the upper left. In the upper right is the T-DNA, which carries genes for indoleacetic acid (tms), cytokinin (tmr), and octopine (ocs) biosynthesis. Elsewhere are genes for octopine catabolism (occ), conjugal transfer (tra), origin of replication, and incompatibility (ori, inc).

Another region of the Ti plasmid, termed the virulence (vir) region, is required for tumor formation but is not found integrated into plant tumor DNA.[8,9] The vir gene products carry out the processing and transfer of T-DNA into plant cells.[10,11]

The unique nature of this infection could, by itself, justify the attention it has received from plant pathologists. However, by evolving a mechanism for active transformation of higher plants, A. tumefaciens has attracted the greatest attention from plant genetic engineers, who use it routinely as a vector for introduction of cloned genes into dicots. For this purpose, several "T-vectors" have been developed. Such vectors carry the vir genes and the portions of the T-DNA which are required for transfer and integration, but are disarmed by removal from the T-DNA of the auxin, cytokinin, and opine biosynthetic genes.[12,13] These genes are replaced with a selectable marker (usually kanamycin resistance under control of an opine biosynthetic promoter), and with the gene or genes which the researcher wishes to introduce into the host. Such vectors are frequently constructed so that the vir genes and the transferred genes are on separate replicons (binary vectors).

T-Vectors are sufficient for some applications, but there are many more for which they are not. Higher efficiencies of transformation of dicots are frequently required, and the host range of A. tumefaciens infection is often restrictive. Some commercially important dicots and most monocots are resistant to transformation by this method.[14,15] For this reason, a complete understanding of the infection as it occurs at the molecular and cellular levels is important.

For convenience, the process of crown gall tumor formation can be divided into five steps, only some of which require the active participation of both bacterium and plant (Table 1). The first step is attachment of the bacteria to sites on the plant cell wall. Second, expression of Ti plasmid vir genes is induced by phenolic compounds produced in wounded plant tissue. Since the inducing compounds are freely diffusible, attachment is not required. Moreover, attachment does not depend upon vir gene induction. The first two steps are therefore independent of each other. Once expressed, some of the vir genes carry out the third step in which the T-DNA is processed to a form which is capable of transfer to the plant cell. This step does not require active participation of plant cell factors. Fourth, by a completely unknown mechanism, the T-DNA is transferred to the plant cell and is then integrated into the plant nuclear DNA. Fifth, the T-DNA genes are expressed, and tumor formation is initiated. This last step is not dependent upon the presence of the bacteria. This report will focus on the bacterial-plant interactions which occur in the first two steps.

Table 1. Steps in crown gall tumor formation

Step	Requirements	
	Bacterial Gene Products	Plant Gene Products
1. Attachment	Bacterial sites	Plant cell wall sites
2. *vir* Gene induction	Positive regulatory proteins (*virA*, *virG*)	Diffusible phenolic compounds
3. T-DNA processing	*vir* Gene products	None
4. T-DNA transfer	*vir* Gene products	Unknown
5. Tumor growth and opine synthesis	None	T-DNA gene products

THE REQUIREMENT FOR ATTACHMENT

A variety of experimental methods has been employed in the study of attachment of bacteria to plant cells in crown gall tumor formation. This has often resulted in a variety of conclusions. However, there is a consensus that attachment is a required step in tumorigenesis, that plant cell walls are the sites of attachment, and that bacterial chromosomal genes code for attachment.

The requirement for attachment of bacteria to plant cells was originally inferred from the observation that avirulent or heat killed strains of *A. tumefaciens*, inoculated onto wounded pinto bean leaves together with virulent strains, inhibit tumor formation by the virulent strains.[16] Similar results were obtained using potato disks[17] and Jerusalem artichoke tuber slices.[18] In these competition experiments, the competing (avirulent) strains are thought to occupy available binding sites on the plants, and thus inhibit attachment and transformation by virulent strains.

In addition to competition experiments *in vivo*, assays are employed in which attachment of bacteria to plant cells in suspension is quantitatively determined *in vitro*. These assays are useful for mutant analysis and for studying the effects of environmental conditions on attachment. However, in addressing the question of a requirement for attachment in tumor formation, they have the limitation that the plant cells which are used in the assays are rarely transformed. An exception is the system, employed by Krens *et al.*,[19] in which transformation of cell-wall-regenerating tobacco protoplasts was found to occur only under conditions in which bacterial attachment to the protoplasts was observed. In separate experiments, an *in vitro* assay for attachment of *Agrobacterium* to root cap cells from 48 plant species has revealed a strong correlation between attachment and susceptibility to tumorigenesis.[20] Moreover, the use of *in vitro* attachment assays in our laboratory and in others has established that bacterial mutants (discussed below) which are defective in attachment to plant cells in suspension are always avirulent. In all of these assays, attachment to plant cells of bacteria such as *Escherichia coli*, which do not interact with plants, is never observed.

PLANT CELL ATTACHMENT SITES

Efforts to identify the sites on the plant cell to which agrobacteria attach have been complicated by the variety of approaches which have been employed. In one report, freshly isolated carrot protoplasts which appear to lack cell wall material were observed to bind *Agrobacterium*.[21] Most of the data, however, support the conclusion that the bacteria attach to pectin-containing sites on the plant cell walls. Plant cell walls, but not cell membranes, inhibit tumor formation on pinto bean leaves,[22] as does polygalacturonic acid.[23] Similar results were obtained with potato disks.[24] Results from *in vitro* assays generally agree with *in vivo* competition experiments. Krens *et al.*[19] concluded that cell wall regeneration is required for attachment and transformation of tobacco protoplasts. Pectin-enriched cell wall fractions strongly inhibit attachment to tomato suspension culture cells,[25] but added pectin does not block attachment of the agrobacteria to carrot suspension culture cells.[26]

There are insufficient data to confirm or deny the intriguing possibility that attachment may be an impediment to transformation of monocots. We have observed specific attachment of agrobacteria to suspension-cultured bamboo cells,[27] but Hawes and Pueppke observed little or no attachment to root cap cells from several monocots.[20] Agrobacterial transformation of monocots in special circumstances has been reported;[28,29] attachment is presumably occurring in these cases.

BACTERIAL CELL ATTACHMENT SITES

Analysis of bacterial attachment mechanisms has been greatly aided in recent years by the ability to carry out mutational analysis and other genetic manipulations on *Agrobacterium*. Early experiments, however, relied primarily upon competition experiments similar to those described above. *Agrobacterium* lipopolysaccharide (LPS) has been implicated in this way since it inhibits tumor formation on pinto bean leaves,[30,31] and blocks attachment of the bacteria to tobacco and carrot tissue culture cells.[32,33] The O-antigen portion of the LPS molecule has been correlated with attachment and virulence in *Agrobacterium*[34,35] and *Pseudomonas solanacearum*.[36,37] However, Pueppke and Benny[24] did not observe inhibition of tumor formation on potato disks by added LPS.

With a few exceptions,[32,38] the bulk of the data we now have suggests that attachment is not encoded by Ti plasmid genes. Avirulent strains lacking the Ti plasmid can compete with virulent strains for attachment sites in tumor inhibition assays,[22,38] and in direct binding assays.[32] Loss of the Ti plasmid does not affect the ability to attach to plant cells *in vitro*.[19,39] In recent years, five chromosomal loci have been identified which are involved in attachment: *chvA*, *chvB*,[40] *exoC*,[41] *cel*,[42] and *att*.[33]

The *chv* and *exo* Loci

We have used both a tumor inhibition assay with Jerusalem artichoke tuber slices, and *in vitro* binding assays consisting of radiolabeled bacteria added to suspension-cultured tobacco cells or to freshly isolated mesophyll cells of Zinnia leaves. By random mutagenesis of the entire *A. tumefaciens* genome with the transposon Tn5, we obtained avirulent, nonattaching mutants.[39,40] Genetic analysis showed that these insertions are clustered within an 11 kilobase chromosomal virulence region carrying two transcriptional units which we designated *chvA* and *chvB*.[40] A genetic map of the region is shown in Figure 2. Earlier mapping data indicated that two insertions that do not result in avirulence mapped within the *chvB* region.[40] However, more recent data shows that these insertions were mismapped (G. Cangelosi, unpublished). Each region, therefore, appears to consist of a single transcriptional unit.

So far, analysis of lipolysaccharide and outer membrane protein content has not revealed any differences between *chvA* and *chvB* mutants and wild type cells (C. Douglas, unpublished). The first clues to the biochemical function of these loci came from analysis of exopolysaccharide synthesis, and of analogous mutations in *Rhizobium meliloti*, a member of the fast-growing Rhizobia which is very closely related to *A. tumefaciens*.[43] The *Rhizobium* loci, termed *ndvA* and *ndvB*, are required for effective nodulation of alfalfa; mutants induce the formation of empty, ineffective nodules which contain no bacteroids and do not fix nitrogen.[44] The *ndv* genes of *Rhizobium* and the *chv* genes of *Agrobacterium* are homologous with each other and are functionally interchangeable in heterologous complementation experiments.

The genera *Agrobacterium* and *Rhizobium* synthesize two unique exopolysaccharides.[45,46,47] One is succinoglycan, a high-molecular weight polysaccharide consisting of repeating octameric units composed of glucose and galactose (7:1) with acidic sidegroups.[45,47] The other is a small (17-20 glucose residues), cyclic β-1,2-glucan.[46] Succinoglycan is excreted into the media in large amounts by both *A. tumefaciens* and *R. meliloti* and is the major constituent of the capsular slime produced by these species. β-1,2-Glucan is primarily a periplasmic polysaccharide, but it

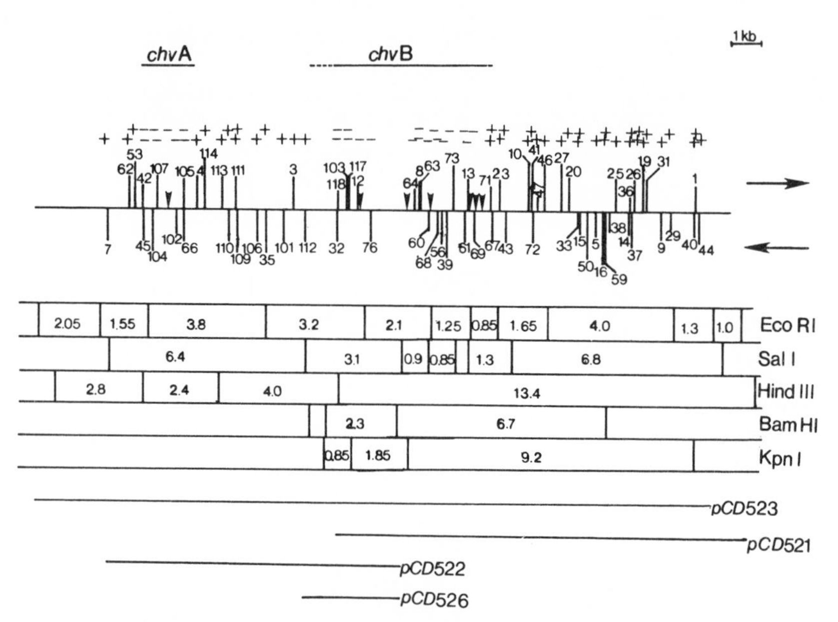

Fig. 2. Genetic map of the chromosomal virulence region, revised from Douglas *et al.* (1985).[40] Positions of transposon insertions are indicated by numbered vertical lines (Tn3::HoHo1 insertions),[40] and vertical arrowheads (Tn5 insertions). The virulence phenotype resulting from each insertion is indicated near the top, insertions resulting in negative phenotypes define the *chvA* and *chvB* loci. By analysis of expression of *chv::lacZ* fusions resulting from Tn3::HoHo1 insertions, the directions of transcription of the *chv* genes were determined. Fusions oriented from left to right (above the line) resulted in detectable *lacZ* expression in the *chvA* region, and those oriented from right to left (below the line) resulted in detectable *lacZ* expression in the *chvB* region.[40] Therefore, the two operons are transcribed in opposite, converging directions. In the middle part of the figure is a restriction map of the region, showing the size of fragments in kilobases. At the bottom are fragments carried by several cosmid clones of the region.[40]

too is excreted by A. tumefaciens and one strain of R. meliloti.[47] Cellular, periplasmic, and extracellular β-1,2-glucan is absent in chvB mutants,[48] and in mutants of R. meliloti which were isolated but not mapped.[49] These mutants of R. meliloti form ineffective nodules similar to ndv mutants, and they may prove to be ndv mutants themselves.

A chvA mutant was reported to synthesize the glucan,[48] but this strain was subsequently found to have a wild-type chvA allele (G. Cangelosi, unpublished). When recent data indicated that neither ndvA nor ndvB mutants synthesize the glucan (G. Ditta, personal communication), we reexamined the exopolysaccharides produced by chvA mutants. When exopolysaccharides from culture supernatants of wild-type A. tumefaciens are separated by size on a Bio-Gel A5-m column, two peaks are observed, corresponding to high molecular weight succinoglycan and low molecular weight β-1,2-glucan.[41,47] The low-molecular weight peak is missing in both chvA and chvB mutant exopolysaccharide (Fig. 3). Thus, it now appears that both chv loci are required for extracellular β-1,2-glucan excretion.

Further studies on β-1,2-glucan production have been carried out by R. Ugalde and coworkers.[50,51] They found that chvB mutants are unable to synthesize β-1,2-glucan in vitro. They also observed a protein-bound intermediate in β-1,2-glucan production in the inner membrane function of wild type but not chvB mutant cells. The protein was estimated to have a molecular weight of around 235 kilodaltons. Recent experiments indicate that chvB contains the structural gene for this peptide (Zorreguita, A., R. Geremia, S. Cavaignac, G.A. Cangelosi, E.W. Nester, and R.A. Ugalde, manuscript submitted to Molecular Plant-Microbe Interactions).

Succinoglycan plays an important role in the invasion of root nodules by R. meliloti. Mutants which do not synthesize this exopolysaccharide form empty, ineffective nodules somewhat similar to those formed by ndv mutants.[52,53] Since β-1,2-glucan appears to play an important role in both Rhizobial and Agrobacterial infections of plants, we were prompted to investigate the possible role of succinoglycan in crown gall tumor formation. Using a fluorescent stain for polysaccharides to screen mutagenized colonies, we isolated succinoglycan-negative (exo) mutants of Agrobacterium analogous to those of Rhizobium.[41] At least five

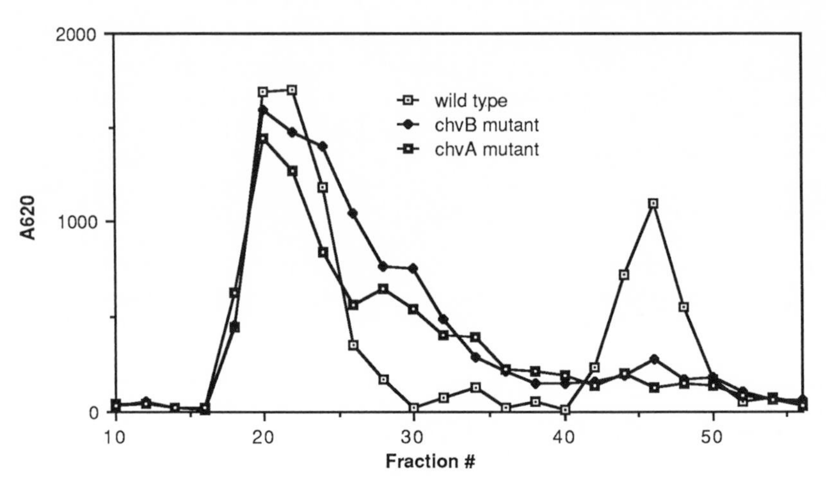

Fig. 3. Size fractionation of wild-type and mutant culture supernatants. Bacteria were grown on a mannitol-glutamate medium, and cells were removed by centrifugation. The culture supernatants were loaded onto a 3 cm X 45 cm BioGel A-5m column, eluted with phosphate buffer, and fractions were assayed for hexose content by anthrone analysis as described previously.[41] By this procedure, high-molecular weight polysaccharide (mostly succinoglycan) elutes first, followed by low-molecular weight polysaccharide (mostly β-1,2-glucan) near the included volu volume.[41,47,48] The wild-type strain was A348, the chvA mutant was ME42 (identical results were obtained with chvA mutant ME66), and the chvB mutant was ME39.[40] The data show that both chvA and chvB mutants produce little or no extracellular, low-molecular weight polysaccharide, compared with the wild-type parent.

different *Agrobacterium* *exo* loci were identified (*exoA*, *exoB*, *exoC*, *exoD*, *exoF*, and *exoG*), and all were found to be functionally interchangeable with those of *Rhizobium* in heterologous complementation experiments. Mutations in any one of these loci eliminates succinoglycan synthesis, but only *exoC* mutants are defective in the synthesis of β-1,2-glucan as well. The *exoC* mutants are also unique among the *exo* mutants in that they are unable to attach to plant cells *in vitro* and form tumors. All of the other *exo* mutants form normal crown gall tumors on a variety of plant

Table 2. Exopolysaccharide production and virulence of Agrobacterium tumefaciens mutants.

Genotype	β-1,2-Glucan Production	Succinoglycan Production	Attachment	Virulence
Wild type	+	+	+	+
chvA or B⁻	-	+	-	-
exoC⁻	-	-	-	-
exoA, B, D, F, or G⁻	+	-	+	+

hosts. These results, combined with the data on the chv mutants and summarized in Table 2, show that β-1,2-glucan is essential for attachment and tumor formation, while succinoglycan is dispensible. For nodulation of alfalfa by Rhizobium, both polysaccharides are required. β-1,2-glucan may be involved in attachment of Rhizobium cells to plant cells. Our results support the suggestion that Rhizobial succinoglycan may be involved in invasion of the nodules by bacterial cells,[53] a process that does not occur in crown gall infection.

The Role of β-1,2-Glucan in Attachment

Having established that β-1,2-glucan plays a role in attachment, it remains for us to identify that role. The cyclic glucans of the Rhizobiaceae resemble the membrane-derived oligosaccharides of the enteric bacteria.[54] It has been suggested that they perform the same function, serving as periplasmic osmotic buffers to enhance tolerance to environments of low osmotic strength. This notion is supported by the observation that synthesis of periplasmic glucans by Agrobacterium and Rhizobium is partially suppressed during growth on high-osmotic strength media.[55] It is possible, therefore, that β-1,2-glucan performs an indirect role in attachment. It may simply be an osmotic stabilizer that is required for the synthesis or function of molecules which are more directly involved. This possibility is supported by the observation that growth of chv mutants on a medium of high osmotic strength will partially restore the ability to attach to tobacco cells (G. Cangelosi, unpublished). However, the ability to form crown gall tumors is not restored by growth on the same medium. Moreover, even when the bacteria are grown on media with salt concentrations high enough to inhibit growth (e.g., 0.5 M NaCl), a fairly high basal level of β-1,2-glucan production is observed.[55] Transcription of the chv genes is not regulated by the osmotic strength of the medium (G. Cangelosi, unpublished). Thus, β-1,2-glucan may play a role in attachment which is independent of its role in osmotolerance.

It is possible that extracellular (as opposed to periplasmic) β-1,2-glucan mediates attachment. This is supported by the fact that R. meliloti strain 1021, which synthesizes β-1,2-glucan but does not excrete it extracellularly like Agrobacterium, does not attach to tobacco cells in vitro (G. Cangelosi, unpublished). Further work to distinguish these possibilities is underway.

The cel and att Loci

A. tumefaciens synthesizes a third major β-linked polysaccharide when it is exposed to plant cells. This polysaccharide is synthesized in the form of fibrils which entangle and trap a large number of bacteria near the surface of the plant cells.[56] The fibrils are probably composed of cellulose and are encoded by chromosomal genes. Mutants which are unable to synthesize them (cel^-) still form crown gall tumors on a variety of plants, but they are somewhat more easily dislodged from inoculated wound sites (by rinsing with water) than are wild-type cells.[42] The cellulose fibrils are therefore thought to perform a secondary role in attachment, as a means of securely anchoring bacteria which have already carried out an initial step in attaching to plant cells. This role appears to be dispensible in laboratory inoculations of wound sites in plants (which generally involve the application of thick suspensions of bacterial cells), but its importance in nature should not be underestimated.

Recently, a new chromosomal attachment locus (att) was identified by the arduous process of screening mutagenized bacteria directly for the ability to attach to suspension-cultured plant cells.[33] The *att* mutants are unable to attach and are avirulent, but they produce normal lipopolysaccharide, β-1,2-glucan, and cellulose fibrils. They differ from the wild type in that they do not produce one or more proteins normally found in the fraction removed from the bacteria during the preparation of spheroplasts (this would presumably include outer membrane and periplasmic proteins). They also exhibit a growth defect on certain media which is severe enough to select for reversion of the transposon mutant phenotype (A. Matthysse, personal communication). Further work is required to identify the missing proteins and to determine their role in attachment and in the general growth of the cell.

CONCLUSIONS ON GENETIC STUDIES OF ATTACHMENT

To date, there are no data to contradict the intuitive assumption that some kind of attachment of bacteria to plant cells is required for tumor formation. Most data now available suggest that the bacteria attach to pectin-containing sites on the plant cell wall. Far more work is needed, however, to confirm that this is the case.

Primarily because of the ease of genetic manipulation of bacteria, we are closer to identifying the bacterial components of attachment. β-1,2-Glucan is directly or indirectly involved, and bacterial cellulose fibrils are probably very important. One very crucial observation is that the Ti plasmid is not required for attachment to plant cells *in vitro*. This observation suggests that attachment may be a more ancestral function than the processing and transfer of T-DNA. Such an assumption would help explain the fact that the *chv* and *exoC* genes do not require plant wound inducers for high levels of expression, as do the Ti plasmid *vir* genes (C. Douglas and G. Cangelosi, unpublished). Moreover, *A. tumefaciens* and *R. meliloti* share the chromosomal attachment loci *chvA*, *chvB*, and *exoC*, all of which are involved in β-1,2-glucan synthesis. Most of the other Rhizobial genes involved in nodulation are borne on plasmids.[57] The very different process of crown gall tumor formation is also primarily plasmid-coded. It is tempting to conclude that the ability to bind plant cells was fully developed in a plant-colonizing ancestor to *Agrobacterium*, and that this ability was simply put to more elaborate use upon the evolution or acquisition of the Ti plasmid.

There are limitations inherent in the study of attachment as a discreet step in tumor formation. We cannot say for certain that the attachment which we observe in our assays in vitro, and which is independent of Ti plasmid genes, is the same attachment which occurs during the transformation of plant cells in wound sites. The vir region covers almost 35 kb of Ti plasmid DNA, and it seems unlikely that none of these genes are involved in cell surface interactions. Such interactions may simply be undetectable in the experimental systems currently in use.

THE VIRULENCE REGION

Attachment is the only step in crown gall tumorigenesis which is primarily encoded by bacterial chromosomal genes. Subsequent steps are catalyzed by proteins which are encoded by Ti plasmid genes. All of the vir genes lie within a 35 kb region of the Ti plasmid near the T-DNA. All are trans-acting factors, i.e., they can catalyze T-DNA transfer even when carried on separate plasmids from the T-DNA, as is the rule in binary plant transformation vectors.[16] A map of the vir region is shown in Figure 4. Four of the vir operons (A,B,D, and G) are absolutely required for transformation, while the loss of the other two (C or E) results in greatly attenuated tumor formation.[58,59,60] To the left of virA is a gene (pinF) which is not required for tumorigenesis, but is coordinately regulated with the vir genes.[61]

Each of the vir loci has been genetically characterized by insertion analysis, restriction mapping, nucleotide sequencing, and genetic complementation experiments. All except virA and virG are polycistronic operons. The protein products of some of the genes, when expressed in both Agrobacterium and E. coli, have been identified. Intensive research is now being conducted to determine the biological function of these gene products, but at this writing only a few have been elucidated. Two of the virD proteins are endonucleases which are directly involved in excision of a single-stranded T-DNA intermediate.[11] One or more of the virE genes appears to act

	INDUCIBILITY	SIZE (Kb)	ORF'S	SIZE (Kd)	MUTANT PHENOTYPE	ROLE	FUNCTION
B	+	9.5	11	-	avir.	Transfer	?
D	+	4.5	4	16 48 21 76	avir.	Transfer	Endonuclease
C	+	1.5	2	26 23	atten.	Accessory	?
E	+	2.2	2	7 60	very atten.	Accessory	?
A	-	2.8	1	92	avir.	Regulation	Phenolic Sensor
G	+	1.0	1	30	avir.	Regulation	Transcriptional Activator

Fig. 4. Map of the virulence region of an octopine Ti plasmid. The position of each vir operon and their directions of transcription are indicated. Courtesy of S. Winans.

extracellularly, since T-DNA transfer from virE$^-$ strains can be "rescued" by coinoculation of wound sites with a virE$^+$ strain.[62] Both virA and virG are directly involved in the positive regulation of vir gene expression in response to plant cell factors. This is the second early step in crown gall tumor formation.

INDUCTION OF VIRULENCE GENE EXPRESSION BY PLANT METABOLITES

The attachment loci encode such products as β-1,2-glucan which is involved in osmoregulation and is therefore useful to the bacteria regardless of plant interactions. The vir genes, on the other hand, appear to function solely in T-DNA transfer. Therefore, it is not surprising that they are tightly regulated so that they are expressed only in the presence of potential plant hosts.

The primary tools for analyzing vir gene expression has been the use of gene fusions in which genes encoding such easily monitored enzymes as β-galactosidase (lac) and chloramphenicol acetyltransferase (cat) are placed under the control of vir promoters. Using vir:lac fusions, expression of virB, virC, virD, virE, and virG was found to be induced 10-300 fold by cocultivation of the bacteria with plant cells, whereas virA expression was unaffected.[61] Partial characterization of the inducing factor showed that it is small (<1000 Da), heat stable, and partially hydrophobic. Subsequently, vir gene induction was observed in the presence of acetosyringone and α-hydroxy-syringone, two phenolic compounds which are released by metabolically active plant cells in response to wounding.[63] Induction by a mixture of plant-derived phenolics has also been observed (Fig. 5).[64] Induction of bacterial genes by plant metabolites has been observed elsewhere: Rhizobium nodulation genes are induced by plant root exudates.[65] In this case, specific flavonoid compounds were found to be the inducing factors.[66] Thus, a superficial resemblance exists between Agrobacterial and Rhizobial infections, in that both bacteria respond to chemical indicators of the presence of appropriate target plant cells.

R_1, R_2, R_3, R_4, $_5R$

COMMON NAME	STRUCTURE	INDUCING ACTIVITY
CATECHOL	R_1=OH; R_2=OH; $R_{3,4,5}$=H	+
P-HYDROXYBENZOIC ACID	R_1=COOH; R_4=OH; $R_{2,3,5}$=H	+
β-RESORCYLIC ACID	R_1=COOH; $R_{2,4}$=OH; $R_{3,5}$=H	+
PROTOCATECHUIC ACID	R_1=COOH; $R_{3,4}$=OH; $R_{2,5}$=H	+ (first seven together: ++++)
PYROGALLIC ACID	R_1=COOH; $R_{2,3,4}$=OH; R_5=H	+
GALLIC ACID	R_1=COOH; $R_{3,4,5}$=OH; R_2=H	+
VANILLIN	R_1=CHO; R_3=OCH_3; R_4=OH; $R_{2,5}$=H	+
ACETOSYRINGONE	R_1=$COCH_3$; $R_{3,5}$=OCH_3; R_4=OH	++++
SYRINGIC ACID	R_1=COOH; $R_{3,5}$=OCH_3; R_4=OH	++++
SINAPINIC ACID	R_1=CHCHCOOH; $R_{3,5}$=OCH_3; R_4=OH	++++

Fig. 5. Plant phenolic compounds which induce expression of the vir genes. Types and positions of ring substitutions are indicated, and the relative ability of each compound to induce the vir genes is represented by plusses. The first seven compounds are most active when present simultaneously.[64] Courtesy of G. Bolton.

Since plant wound phenolics are freely diffusible, attachment to plant cells should not be necessary for induction. In cocultivation experiments, unattached bacteria (those easily separated from plant cells by filtration) contain fully induced *vir* genes, and of course no attachment is possible when the *vir* genes are induced by phenolic compounds in the absence of plant cells. Conversely, attachment does not require any Ti plasmid genes, which would include the two *vir* genes (A and G) that mediate regulation by plant metabolites. Therefore, attachment and *vir* gene induction should not be considered sequential steps but rather independent and probably simultaneous early events in tumorigenesis.

Induction of the virulence genes is another level at which host range of *Agrobacterium* could be determined. Recent work has shown that Agrobacterial transformation of yam (a monocot) is possible when the bacteria are preinduced with wound exudates from potato plants.[29] The inability to infect yam under normal circumstances, therefore, appears to be due to the absence of suitable inducing compounds. We have observed that cultured rice cells are not only unable to induce virulence gene expression, but that they actually antagonize the ability of added acetosyringone to carry out the induction (A. Tanaka, unpublished). These observations suggest that virulence gene regulation could be a limiting factor in *Agrobacterium* host range.

THE ROLE OF *virA* AND *virG* IN VIRULENCE GENE INDUCTION

To determine whether any of the *vir* genes mediate regulation by plant cell factors, *vir-lac* fusions were again employed. It was observed that both *virA* and *virG* are required for induction of the rest of the *vir* genes in both cocultivation experiments and in the presence of phenolic inducers.[67,68]

Computer analysis of *virA* and *virG* nucleotide sequences revealed that the two genes are strongly homologous with genes involved in several other two-component bacterial regulatory systems.[68,69] The regulatory targets of these systems are diverse, but the systems share the common function of sensing environmental signals and carrying out appropriate regulatory responses. *virA* and *virG* are homol-

Table 3. Two-component bacterial regulatory systems homologous to virA and virG.

Homologous to virA	Homologous to virG	System Regulated
envZ	ompR	outer membrane proteins
ntrB	ntrC	nitrogen-regulated genes
phoR	phoB	phosphate-regulated genes
cpxA	sfrA	F transfer genes, etc.

ogous to, respectively, envZ and ompR, which regulate the synthesis of outer membrane proteins in response to osmotic conditions; ntrB and ntrC, which regulate nitrogen assimilation genes in response to availability of nitrogen; phoR and phoB, which respond to phosphate starvation; and cpxA and sfrA, which regulate the expression of F plasmid transfer genes as well as a number of inner membrane proteins. These relationships, summarized in Table 3, show that virA and virG belong to a family of two-component regulatory systems. Both virA and virG are also homologous to several other regulatory proteins.

Computer analysis of nucleotide sequences revealed additional facts about the virA and virG proteins. The virG protein contains a large number of charged amino acids and no significant hydrophobic domains, suggesting that it is a soluble protein.[68] The virA protein, on the other hand, contains two significant hydrophobic domains, suggesting that the protein may be inserted in the cell membrane with hydrophilic portions exposed on either side of the membrane.[69] Both hydrophobic domains are followed by a series of positively charged amino acids. This pattern is typical of the "signal sequences" found in a number of transmembrane proteins.[70] To confirm a membrane localization for the virA protein, antiserum raised against the protein was used as a probe in immunoblotting experiments. Whole Agrobacterium cells were fractionated into outer

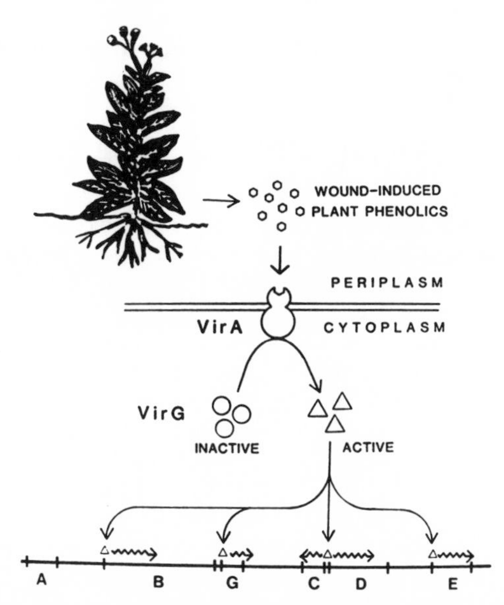

Fig. 6. Model for regulation of vir gene expression by plant wound phenolics (adapted from Winans et al., 1986).[68] See text for explanation.

membrane, inner membrane, and cytoplasmic fractions, and the virA protein was observed only in the inner membrane fraction.[69]

These data, combined with data already available on homologous two-component regulatory systems, lead to a model for regulation of the virulence genes by plant phenolics (Fig. 6). The virA protein is inserted in the inner membrane with portions of the protein exposed to both the exterior and interior of the cell. The exterior portion of the protein detects the environmental signal, in this case the plant phenolics. Interestingly, this portion of the protein shows the greatest divergence in sequence from corresponding portions of homologous regulatory proteins. These may be the regions which interact with unique environmental signals. Upon sensing the inducer, the cytoplasmic portion of the virA protein interacts with the virG protein. By analogy with homologous systems, this interaction may take the form of conversion of the virG protein from an inactive to an active form. The active virG protein then promotes transcription of the vir genes.

CHROMOSOMAL REGULATION OF THE VIRULENCE GENES

In an attempt to identify genes involved in regulation of the vir genes, Close et al.[71] used chloramphenicol to select for mutants which derepress a virC:cat fusion under conditions which normally repress virC. The mutants which they isolated in this fashion were also found to be derepressed for virD, but not for the other vir genes. The mutation was mapped to the bacterial chromosome. Pleiotrophic effects of the chromosomal mutation include repression of exopolysaccharide synthesis and temperature sensitive growth. The new locus was termed ros, for rough surface, because the exopolysaccharide defect results in an altered colony morphology. These data suggest that virC and virD, while coregulated with the other vir genes by virA, virG and plant phenolics, are also controlled by a more general, chromosomal regulatory system. In the presence of plant inducers, virC and virD are expressed to nearly normal induced levels in ros mutants, and tumorigenesis does not seem to be affected.

CONCLUSIONS

The bacteria are the better understood of the two partners involved in crown gall initiation, due to the relative ease of manipulating prokaryotes. The two initial steps in transformation of dicotyledonous plants by A. tumefaciens which were discussed in this report are attachment of bacteria to plant cells, and induction of bacterial virulence genes by plant-derived phenolic compounds. Close examination of these processes is revealing a common feature. Both appear to be adaptations of physiological functions which are not unique to Agrobacteria. For example, attachment is encoded by at least three chromosomal loci (chvA, chvB, and exoC) which are also found in the fast-growing Rhizobia, and further examination may reveal that these genes are present in other plant-colonizing bacteria as well. These three genes code for the synthesis of β-1,2-glucan which may play a completely unrelated role in osmoregulation. Induction of the vir genes is mediated by two genes (virA and virG) which are closely homologous to a variety of functionally unrelated, two-component, regulatory systems found in a diverse range of bacteria. Two of the vir genes are also regulated by a chromosomal locus (ros).

The third step in the transformation process, that is T-DNA processing, may bear some similarity to the processing which occurs during the conjugational transfer of plasmid DNA between bacteria. The late steps in tumorigenesis are transfer, integration, and expression of bacterial DNA in plant cells, and these still appear to be completely unique in nature. Further research, however, may reveal that these steps also have similarities (and perhaps common evolutionary roots) with more familiar phenomena. Nonetheless, each step in crown gall tumor formation requires adaptations which, taken together, make transformation of plants by Agrobacterium a truly unique phenomenon. A thorough understanding of these adaptations may broaden the usefulness of this phenomenon in plant biotechnology.

ACKNOWLEDGMENTS

We thank Gary Ditta for sharing with us his unpublished data on β-1,2-glucan synthesis by Rhizobium mutants. Experiments conducted by coworkers in our laboratory and in other laboratories around the world are cited; their contributions to the field are appreciated. G.A.C. is supported by an American Cancer Society fellowship, and other work in our laboratory is supported by grants from the National Science Foundation, the National Institutes of Health, and the U.S. Department of Agriculture.

REFERENCES

1. CHILTON, M.-D., M.H. DRUMMOND, D.J. MERLO, D. SKIAKY, A.L. MONTOYA, M.P. GORDON, E.W. NESTER. 1977. Stable incorporation of plasmid DNA into higher cells: the molecular basis of crown gall tumorigenesis. Cell 11: 263-271.
2. THOMASHOW, M.F., R. NUTTER, A.L. MONTAGU, M.P. GORDON, E.W. NESTER. 1980. Integration and organization of Ti-plasmid sequences in crown gall tumors. Cell 19: 729-739.

3. ZAMBRYSKI, P., M. HOLSTERS, K. KRUGER, A. KEPICKER, J. SCHELL, M. VAN MONTAGU, H.M. GOODMAN. 1980. Tumor DNA structure in plant cells transformed by A. tumefaciens. Science 209: 1385-1391.
4. AKIYOSHI, D.E., R.O. MORRIS, R. HINZ, B.S. MISCHKE, T. KOSUGE, D.J. GARFINKEL, M.P. GORDON, E.W. NESTER. 1982. Cytokinin/auxin balance in crown gall tumors is regulated by specific loci in the T-DNA. Proc. Natl. Acad. Sci. USA 80: 407-411.
5. AMASINO, R.M., C.O. MILLER. 1982. Hormonal control of tobacco crown gall tumor morphology. Plant Physiol. 69: 389-392.
6. WILLMITZER, L., G. SIMONS, J. SCHELL. 1982. The T-DNA in octupine crown gall codes for seven well-defined polyadenylated transcripts. EMBO J. 1: 139-146.
7. BOMHOFF, G.H., P.M. KLAPWIJK, H.C.M. KESTER, R.A. SCHILPEROORT, J.P. HERNALSTEENS, J. SCHELL. 1976. Octopine and nopaline synthesis and breakdown is genetically controlled by a plasmid of Agrobacterium tumefaciens. Mol. Gen. Genet. 145: 177-181.
8. KLEE, H.J., F.F. WHITE, V.N. IYER, M.P. GORDON, E.W. NESTER. 1983. Mutational analysis of the vir region of an Agrobacterium tumefaciens Ti plasmid. J. Bacteriol. 153: 878-883.
9. LUNDQUIST, R.C., T.J. CLOSE, C.I. KADO. 1984. Genetic complementation of Agrobacterium tumefaciens Ti plasmid mutants in the virulence region. Mol. Gen. Genet. 193: 1-7.
10. HORSCH, R.B., H.J. KLEE, S. STACHEL, S.C. WINANS, E.W. NESTER, S.G. ROGERS, R.T. FRALEY. 1986. Analysis of Agrobacterium tumefaciens virulence mutants in leaf discs. Proc. Natl. Acad. Sci. USA 82: 2571-2575.
11. YANOFSKY, M.F., S. PORTER, C. YOUNG, L. ALBRIGHT, M.P. GORDON, E.W. NESTER. 1986. The virD operon of Agrobacterium tumefaciens encodes a site-specific endonuclease. Cell 47: 471-477.
12. AN, G., B.D. WATSON, S. STACHEL, M.P. GORDON, E.W. NESTER. 1985. New cloning vehicles for transformation of higher plants. EMBO J. 4: 277-284.
13. ZAMBRYSKI, P., H. JOOS, C. GENETELLO, J. LEEMANS, M. VAN MONTAGU, J. SCHELL. 1983. Ti plasmid vector for the introduction of DNA into plant cells without alteration of their normal regeneration capacity. EMBO J. 2: 2143-2150.

14. DE CLEENE, M., J. DE LEY. 1976. The host range of crown gall. Bot. Rev. 42: 389-466.
15. DE CLEENE, M. 1985. The susceptibility of monocotyledons to Agrobacterium tumefaciens. Phytopath. Z. 113: 81-89.
16. LIPPINCOTT, B.B., J.A. LIPPINCOTT. 1969. Bacterial attachment to a specific wound site as an essential stage in tumor initiation by Agrobacterium tumefaciens. J. Bacteriol. 97: 620-628.
17. GLOGOWSKI, W., A.G. GALSKY. 1978. Agrobacterium tumefaciens site attachment as a necessary prerequisite for crown gall tumor formation of potato discs. Plant Physiol. 61: 1031-1033.
18. TANIMOTO, E., C. DOUGLAS, W. HALPERIN. 1979. Factors affecting crown gall tumorigenesis in Jerusalem artichoke (Helianthus tuberosum L.). Plant Physiol. 63: 989-994.
19. KRENS, F.A., L. MOLENDIJK, G.J. WULLEMS, R.A. SCHILPEROORT. 1985. The role of bacterial attachment in the transformation of cell-wall-regenerating tobacco protoplasts by Agrobacterium tumefaciens. Planta 166: 300-308.
20. HAWES, M.C., S.G. PUEPPKE. 1986. Correlation between binding of Agrobacterium tumefaciens by root cap cells and susceptibility of plants to crown gall. Plant Cell Reports, in press.
21. MATTHYSSE, A.G., K.V. HOLMES, R.H.G. GURLITZ. 1982. Binding of Agrobacterium tumefaciens to carrot protoplasts. Physiol. Plant Pathol. 20: 27-33.
22. LIPPINCOTT, J.A., B.B. LIPPINCOTT. 1977. Tumor induction by Agrobacterium involves attachment of the bacterium to a site on the host plant cell wall. Plant Physiol. 59: 388-390.
23. RAO, S.S., B.B. LIPPINCOTT, J.A. LIPPINCOTT. 1982. Agrobacterium adherence involves the pectin portion of the host cell wall and is sensitive to the degree of pectin methylation. Physiol Plant Pathol. 56: 374-380.
24. PUEPPKE, S.G., U.K. BENNY. 1983. Agrobacterium tumorigenesis in potato: effect of added Agrobacterium lipopolysaccharides and the degree of methylation of added plant galacturonase. Physiol. Plant Pathol. 23: 439-446.
25. NEFF, N.T., A.N. BINNS. 1985. Agrobacterium tumefaciens interaction with suspension-cultures tomato

cells. Plant Physiol. 77: 35-42.
26. MATTHYSSE, A.G., R.H.G. GURLITZ. 1982. Plant cell range for attachment of *Agrobacterium tumefaciens* to tissue culture cells. Physiol. Plant Pathol. 21: 381-387.
27. DOUGLAS, C., W. HALPERIN, M. GORDON, E.W. NESTER. 1985. Specific attachment of *Agrobacterium tumefaciens* to bamboo cells in suspension cultures. J. Bacteriol. 161: 764-766.
28. HOOYKASS-VAN SLOGTEREN, G.M.S., P.J. HOOKAAS, R.A. SCHILPEROORT. 1984. Expression of Ti plasmid genes in monocotyledonous plants infected with *Agrobacterium tumefaciens*. Nature 311: 763-764.
29. SCHAFER, W., A. GORZ, G. KAHL. 1987. T-DNA integration and expression in a monocot crop plant after induction of *Agrobacterium*. Nature 327: 529-532.
30. WHATLEY, M.H., J.S. BODWIN, B.B. LIPPINCOTT, J.A. LIPPINCOTT. 1976. Role for *Agrobacterium* cell envelope lipopolysaccharide in infection site attachment. Inf. Immunol. 13: 1080-1083.
31. NEW, D.B., J.J. SCOTT, C.R. IRELAND, S.K. FARRAND, B.B. LIPPINCOTT, J.A. LIPPINCOTT. 1983. Plasmid pSa causes loss of LPS-mediated adherence in *Agrobacterium*. J. Gen. Microbiol. 129: 3657-3660.
32. MATTHYSSE, A.G., P.M. WYMAN, K.V. HOLMES. 1978. Plasmid dependent attachment of *Agrobacterium tumefaciens* to plant tissue culture cells. Inf. Immunol. 22: 516-522.
33. MATTHYSSE, A.G. 1987. Characterization of non-attaching mutants of *Agrobacterium tumefaciens*. J. Bacteriol. 169: 313-323.
34. WHATLEY, M.H., B.B. LIPPINCOTT, J.A. LIPPINCOTT. 1976. Site attachment in *Agrobacterium* infection involves bacterial "O-antigen". Amer. Soc. Microbiol. Absts. Ann. Meeting, B8.
35. BANERJEE, D., M. BASA, I. CHOUDHURY, G.C. CHATERJEE. 1981. Cell surface carbohydrates of *Agrobacterium tumefaciens* involved in adherence during crown gall tumor initiation. Biochem. Biophys. Res. Commun. 100: 1348-1388.
36. WHATLEY, M.H., N. HUNTER, M.A. CANTRELL, C. HENDRICK, K. KEEGSTRA, L. SEQUEIRA. 1980. Lipopolysaccharide composition of the wilt pathogen, *Pseudomonas*

solanacearum: correlation with hypersensitive response in tobacco. Plant Physiol. 65: 557-559.
37. BAKER, C.J., M.J. NEILSON, L. SEQUEIRA, K.G. KEEGSTRA. 1984. Chemical characterization of the lipopolysaccharide of Pseudomonas solanacearum. Appl. Environ. Microbiol. 47: 1096-1100.
38. WHATLEY, M.H., J.B. MARGOT, J. SCHELL, B.B. LIPPINCOTT, J.A. LIPPINCOTT. 1978. Plasmid and chromosomal determination of Agrobacterium adherence specificity. J. Gen. Microbiol. 107: 395-398.
39. DOUGLAS, C.J., W. HALPERIN, E.W. NESTER. 1982. Agrobacterium tumefaciens mutants affected in attachment to plant cells. J. Bacteriol. 152: 1265-1275.
40. DOUGLAS, C.J., R.J. STANELONI, R.A. RUBIN, E.W. NESTER. 1985. Identification and genetic analysis of an Agrobacterium tumefaciens chromosomal virulence region. J. Bacteriol. 161: 850-860.
41. CANGELOSI, G.A., L. HUNG, V. PUVANESARAJAH, G. STACEY, D.A. OZGA, J.A. LEIGH, E.W. NESTER. 1987. Common loci for Agrobacterium tumefaciens and Rhizobium meliloti exopolysaccharide synthesis and their roles in plant interactions. J. Bacteriol. 169: 2086-2091.
42. MATTHYSSE, A.G. 1983. Role of bacterial cellulose fibrils in Agrobacterium tumefaciens infection. J. Bacteriol. 154: 906-915.
43. KRIEG, N.R., J.G. HOLT, eds. 1984. Bergey's Manual of Systematic Bacteriology. Vol. 1, Williams and Wilkins, Baltimore, Maryland.
44. DYLAN, T., L. IELPI, S. STANFIELD, L. KASHYAP, C. DOUGLAS, M. YANOFSKY, E. NESTER, D.R. HELINSKI, G. DITTA. 1986. Rhizobium meliloti genes required for nodule development are related to chromosomal virulence genes in Agrobacterium tumefaciens. Proc. Natl. Acad. Sci. USA 83: 4403-4407.
45. HISAMATSU, M., J. ABE, A. AMEMURA, T. HARADA. 1980. Structural elucidation of succinoglycan and related polysaccharides from Agrobacterium and Rhizobium by fragmentation with two special β-D-glycanases and methylation analysis. Agric. Biol. Chem. 44: 1049-1055.
46. DELL, A., W.S. YORK, M. McNEIL, A.G. DARVIL, P. ALBERSHEIM. 1983. The cyclic structure of β-D-

(1,2)-linked D-glucans secreted by Rhizobia and Agrobacteria. Carbohydr. Res. 117: 185-200.

47. ZEVENHUIZEN, L.P.T.M., A.R.W. VAN NEERVEN. 1983. (1,2)-β-D-glucan and acidic oligosaccharides produced by Rhizobium meliloti. Carbohydr. Res. 118: 127-134.

48. PUVANESARAJAH, V., F.M. SCHELL, G. STACEY, C.J. DOUGLAS, E.W. NESTER. 1985. A role for 2-linked-β-D-glucan in the virulence of Agrobacterium tumefaciens. J. Bacteriol. 164: 102-106.

49. GEREMIA, R.A., S. CAVAIGNAC, A. ZORREGUITA, N. TORO, J. OLIVARES, R.A. UGALDE. 1987. A Rhizobium meliloti mutant that forms ineffective pseudo-nodules in alfalfa produces exopolysaccharide but fails to form β-(1,2) glucan. J. Bacteriol. 169: 880-884.

50. ZORREGUIETA, A., R.A. UGALDE. 1986. Formation in Rhizobium and Agrobacterium spp. of a 235-kilodalton protein intermediate in β-D-(1,2) glucan synthesis. J. Bacteriol. 167: 947-951.

51. ZORREGUIETA, A., R.A. UGALDE, L.F. LeLOIR. 1985. An intermediate in cyclic 1,2-β-glucan biosynthesis. Biochem. Biophys. Res. Commun. 126: 352-357.

52. FINAN, T.M., A.M. HIRSCH, J.A. LEIGH, E. JOHANSEN, G.A. KULDAU, S. DEEGAN, G.C. WALKER, E.R. SIGNER. 1985. Symbiotic mutants of Rhizobium meliloti that uncouple plant from bacterial differentiation. Cell 40: 869-877.

53. LEIGH, J.A., E.R. SIGNER, G.C. WALKER. 1985. Exopolysaccharide-deficient mutants of Rhizobium meliloti that form ineffective nodules. Proc. Natl. Acad. Sci. USA 82: 6231-6235.

54. KENNEDY, E.P. 1982. Osmotic regulation and biosynthesis of membrane-derived oligosaccharides in Escherichia coli. Proc. Natl. Acad. Sci. USA 79: 1092-1095.

55. MILLER, K.J., E.P. KENNEDY, V.N. REINHOLD. 1986. Osmotic adaptation by gram-negative bacteria: possible role for periplasmic oligosaccharides. Science 231: 48-51.

56. MATTHYSSE, A.G., K.V. HOLMES, R.H.G. GURLITZ. 1981. Elaboration of cellulose fibrils by Agrobacterium tumefaciens during attachment to carrot cells. J. Bacteriol. 145: 583-595.

57. LONG, S.R. 1984. Genetics of Rhizobium nodulation. In Plant-Microbe Interactions: Molecular and

Genetic Perspectives. (T. Kosuge, E.W. Nester, eds.), Vol. 1, Macmillan, New York, pp. 265-306.

58. HOOYKAAS, P.J.J., M. HOFKER, H. DEN DULK-RAS, R.A. SCHILPEROORT. 1984. A comparison of virulence determinant in an octopine Ti plasmid, a nopaline Ti plasmid and an Ri plasmid by complementation analysis of Agrobacterium tumefaciens mutants. Plasmid 11: 195-205.

59. KLEE, H.J., F.F. WHITE, V.N. IYER, M.P. GORDON, E.W. NESTER. 1983. Mutational analysis of the vir region of an Agrobacterium tumefaciens Ti plasmid. J. Bacteriol. 153: 878-883.

60. STACHEL, S.E., E.W. NESTER. 1986. The genetic and transcriptional organization of the vir region of the A6 Ti plasmid of Agrobacterium tumefaciens. EMBO J. 5: 1445-1454.

61. STACHEL, S.E., E.W. NESTER, P. ZAMBRYSKI. 1986. A plant cell factor induced Agrobacterium tumefaciens vir gene expression. Proc. Natl. Acad. Sci. USA 83: 379-383.

62. OTTEN, L., H. DE GREVE, J. LEEMANS, R. HAIN, P. HOOYKAAS, J. SCHELL. 1984. Restoration of virulence of Vir region mutants of Agrobacterium tumefaciens strain B6S3 by coinfection with normal and mutant Agrobacterium strains. Mol. Gen. Genet. 195: 159-163.

63. STACHEL, S.E., E. MESSENS, M. VAN MONTAGU, P. ZAMBRYSKI. 1985. Identification of the signal molecules produced by wounded plant cells that activate T-DNA transfer in Agrobacterium tumefaciens. Nature 318: 624-629.

64. BOLTON, G.W., E.W. NESTER, M.P. GORDON. 1986. Plant phenolic compounds induce expression of the Agrobacterium tumefaciens loci needed for virulence. Science 232: 983-985.

65. MULLIGAN, J., S. LONG. 1985. Induction of Rhizobium meliloti nodC expression by plant exudate requires nodD. Proc. Natl. Acad. Sci. USA 82: 6609-6613.

66. PETERS, N.K., J.W. FROST, S.R. LONG. 1986. A plant flavone, luteolin, induces expression of Rhizobium meliloti nodulation genes. Science 233: 977-980.

67. STACHEL, S.E., P.C. ZAMBRYSKI. 1986. virA and virG control the plant-induced activation of the T-DNA transfer process of Agrobacterium tumefaciens. Cell 46: 325-333.

68. WINANS, S.C., P.R. EBERT, S.E. STACHEL, M.P. GORDON, E.W. NESTER. 1986. A gene essential for Agrobacterium virulence is homologous with a family of positive regulatory loci. Proc. Natl. Acad. Sci. USA 83: 8278-8282.
69. LEROUX, B., M.F. YANOFSKY, S.C. WINANS, J.E. WARD, S.F. ZIEGLER, E.W. NESTER. 1987. Characterization of the virA locus of Agrobacterium tumefaciens: a transcriptional regulator and host range determinant. EMBO J. 6: 849-856.
70. WICKNER, T., H.F. LODISH. 1985. Multiple mechanisms of protein insertion into membranes. Science 230: 400-407.
71. CLOSE, T.J., R.C. TAIT, C.I. KADO. 1985. Regulation of Ti plasmid virulence genes by a chromosomal locus of Agrobacterium tumefaciens. J. Bacteriol. 164: 774-781.

Chapter Seven

PLANT STRESS RESPONSES: DISCUSSION OF MODELS FOR RACE-SPECIFIC RESISTANCE

DAVID N. KUHN

Department of Biochemistry
Purdue University
West Lafayette, Indiana 47907

INTRODUCTION

Plants can respond actively to changes in their environment as well as resisting changes by constitutive barriers such as the cuticle and cell wall. In general, plants are in a dynamic equilibrium with the environment, constantly responding at the organ, tissue, cellular as well as the transcriptional, translational and enzyme level to environmental changes. When these changes are extreme or when the environment damages or weakens the plant's ability to grow and reproduce, we term this stress. Biotic stress comes from another living organism (bacterial or fungal pathogen, insect or herbivore). Abiotic stress is caused by environmental extremes (heat, cold, water) or compounds present at toxic levels (heavy metals, ozone).

We are able to study stress responses because the response is usually large and easy to measure, the duration and amount of stress can be artificially manipulated, and large numbers of plants can be placed under identical stress conditions to reduce biological variability. We tend to narrow our focus to the parameters by which we measure the response, usually manifested by secondary metabolite production. Plants make a variety of secondary metabolites in response to stress. We are unaware of the functions of most of these metabolites. Metabolite production requires the transcription of particular genes, translation into protein and the regulation of the newly synthesized enzymes by compartmentation, substrate supply, covalent/noncovalent modification, and degradation. These subcellular responses are easier to quantitate than responses at the tissue or organ level. However, this narrow focus makes transcriptional changes or changes in enzyme activation seem more important than other, more complex parts of the response. Also, researchers often try to isolate a particular factor and, mistakenly, to demonstrate that it is the sole cause for a response. It is perhaps due to these simplifications of our study of a complex response that our models of plant response are also very simple. We are perhaps artificially limiting our view of the response to parameters that can be quantitated. We may tend to ignore or reject less quantifiable data that do not fit the model rather than expanding the model to explain all the data. This may slow our progress in unravelling the complexity of the plant response.

Progress on a basic understanding of plant response is also slowed by research on so many different plants and their response to many different stresses. One approach is to study the response of a few representative plants (monocot, dicot, gymnosperm) to a few representative stresses. Given the extensive diversity of plants, however, it would be difficult if not impossible to generalize from the results. A better approach is to study in great detail and at as many levels as possible one plant's response to one type of stress. In this paper, we discuss the response of legumes, particularly soybeans, to pathogen stress. We first discuss why pathogen stress in soybeans is an excellent system for research. Next, data on the differential response of soybean leaves and roots to pathogens are presented. These data are discussed with regards to the present model of the plant-pathogen interaction and an alternate model is proposed. Finally, some directions for future research on the plant-pathogen interaction that come from consideration of the alternate model are proposed.

ADVANTAGES OF STUDYING PATHOGEN STRESS OF LEGUMES

Legumes make excellent representative plants for research studies. They include important crop plants such as soybeans. Various cultivars with defined pathogen resistance genotypes are commercially available. Large numbers of plants can be grown in fields seasonally or in greenhouses or controlled environment chambers on a year-round basis. Disease resistance testing is not limited to mature plants as young plants (2-10 days old) exhibit the same biochemical responses characteristic of resistance in the mature plants.

A large body of information on phytoalexin biosynthesis exists for legumes.[1,2] In soybeans, pterocarpan phytoalexins (glyceollin I, II, III) are produced in response to pathogen attack. The proposed pathway for glyceollin biosynthesis is presented in Figure 1.[1,3-5] Molecular probes and genes are available for some of the phytoalexin biosynthetic enzymes such as phenylalanine ammonia-lyase[6,7] (PAL) and chalcone synthase[8,9] (CHS) shown in Figure 1. Therefore, characterization of transcriptional and translational changes after inoculation with a pathogen are possible.

Soybean is infected by both airborne and soilborne pathogens and therefore the response in the aerial and subterranean organs can be studied. The soybean response to pathogens is complex and occurs at the subcellular, cellular, tissue and organ levels. Some examples of this response are ethylene synthesis[10] and the accumulation of the phytoalexin glyceollin at the subcellular level,[11,12] strengthening of cell walls at the cellular level,[10,13] the rapid cell necrosis at the site of infection or hypersensitive response at the tissue level,[14] and abscission of infected leaves or the increase of secondary root growth in infected roots at the organ level.

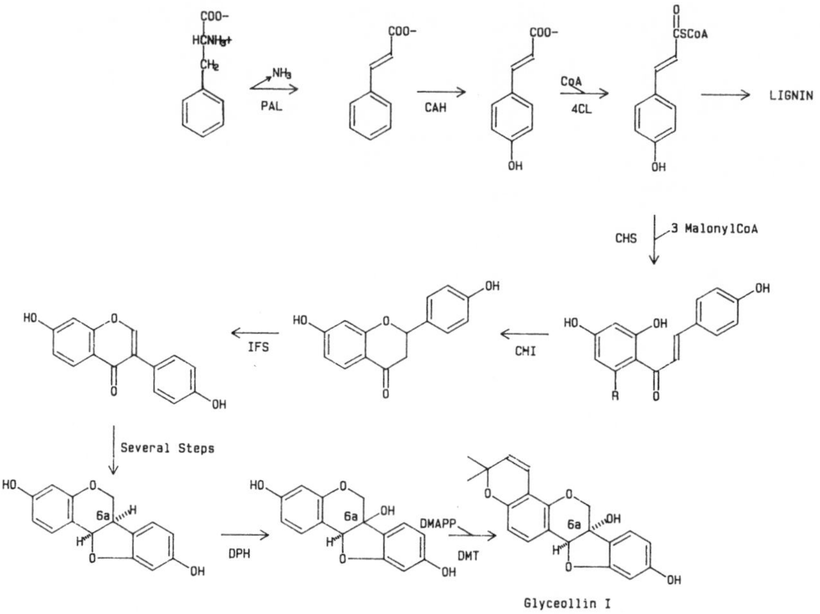

Fig. 1. Proposed glyceollin biosynthetic pathway. Abbreviations: PAL, phenylalanine ammonia-lyase; CAH, cinnamic acid hydroxylase; 4CL, 4-coumarate:CoA ligase; CHS, chalcone synthase; CHI, chalcone isomerase; IFS, isoflavone synthase; DPH, dihydroxypterocarpan 6a-hydroxylase; DMT, dimethylallylpyrophosphate: dihydroxypterocarpan 6a-hydroxyl transferase. For tetrahydroxy chalcone, R = OH; for trihydroxy chalcone, R = H, DMAPP = dimethylallyl pyrophosphate.

There are two possible outcomes for plant-pathogen interactions: resistance (incompatibility) and susceptibility (compatibility). The resistant plant recognizes the pathogen and activates a defense response that limits the pathogen's growth. Phenotypically, the plant will have small lesions or necrotic flecks but the majority of the tissue will be healthy. The susceptible plant does not recognize the pathogen and either does not activate a defense response or activates it too late to limit the pathogen's growth. As the pathogen grows, the plant cells are colonized or killed; this process appears as a spreading of chlorosis in leaves or gives a watersoaked, soft appearance to the tissue that is distinctly different from the necrotic tissue in the resistant plant.

An unusual aspect of the plant response to pathogens is the hypersensitive response, a rapid, plant-directed necrosis of the plant cells near the infection site that limits the growth of the pathogen.[14] The hypersensitive response is easily recognized at the macroscopic level, but it is difficult to quantify and is often overlooked as an indicator of the plant response. Whereas plants and animals respond similarly to heat shock and to heavy metal stress,[15] the hypersensitive response, where some cells in a tissue are sacrificed (by unknown means) to save the whole organism, is unique to plants. Inflammation in animals also involves cell sacrifice, but the animal cells are specialized for that purpose. The hypersensitive response occurs in plants whether phytoalexins are produced or not. Therefore, this type of response is not necessarily due to production of a toxic substance by the plant. The biochemical basis of the hypersensitive response is unknown. We suggest here that the hypersensitive response may be the result of a synergism among several plant stress responses. It may represent the extreme end of a spectrum of synergistic responses where the other end is susceptible.

Pathogen stress is also unique because the plant response must be quickly induced and be rapidly suppressed to prevent cell necrosis from spreading to surrounding tissues. In response to heavy metals, plants accumulate phytochelatins;[16] since high concentrations of phytochelatins are not toxic to the cells, there is no need to have a special mechanism to suppress the response. In response to leaf wounding, some plants accumulate proteinase inhibitors[17] in leaves distant from the wound and continued accumulation is not damaging to the cell. When the plant responds to a pathogen with cell death, it cannot afford to have all of the cells in the attacked organ die, or to have cells at a distance from the infection site die.

Some pathogens develop subpopulations (races) that infect certain cultivars of a plant species but do not infect others. The cultivars that are resistant to certain races of a pathogen (but not all) have race-specific resistance. Not all plant pathogens have races, nor do all plants have race-specific resistance. Race-specific resistance and the interaction of races and cultivars has been extensively studied by Flor,[18] who described some general aspects of the interaction in his gene for gene hypothesis.

Flor's hypothesis was developed to explain genetic data on flax and flax rust. Flax rust is an obligate parasite on flax and is spread from plant to plant by airborne dispersal of an asexual spore. Isolates of flax rust could be differentiated by their ability to successfully infect particular cultivars of flax and their inability to infect other cultivars of flax. The flax rust isolates were thus grouped by their pathogenicity phenotype into pathogenic races and the cultivars grouped by race-specific resistance phenotype. Flor's interest was in the inheritance of race-specific resistance in the plant and virulence (cultivar-specific pathogenicity) in the pathogen. Flor observed that: 1. Race-specific resistance is due to single dominant genes in the host. 2. Virulence is a recessive trait in the pathogen. Avirulence is dominant. 3. If the cultivar has a dominant resistance gene (R1) and the pathogen has a dominant avirulence gene (A1), resistance (incompatibility) results. All other combinations give susceptibility (compatibility) (see Table 1).

Flor's hypothesis is a genetic explanation of genetic data. What we are seeking now is to understand at the biochemical level how Flor's hypothesis functions. What do the products of a resistance gene and an avirulence gene do and how does this lead to race-specific resistance?

Flor's research is important to our study of plant responses because the response is controlled by a single gene. We are certain that many genes are activated in the race-specific resistance response in legumes. The plant response to heat shock, salt stress, heavy metals, etc. also requires the activation of many genes.[15] Only in race-specific resistance do we have genetic evidence that the entire response is controlled by a single gene. This has two important effects. First, one can incite either resistance or susceptibility in a single cultivar with closely related pathogens and thus characterize the differences in response that lead to resistance. Second, when a resistance gene is isolated, we will have the means to probe the essential steps of race-specific resistance.

Table 1. Host-Pathogen Interaction Phenotypes in Flor's Gene-For-Gene Hypothesis.

	Pathogen Genotype	
Host Genotype	Race 1 (A1,A1)[b]	Race 2 (a1,a1)
Cultivar 1 (R1,R1)[a]	R/I[c]	S/C[d]
Cultivar 2 (r1,r1)	S/C	S/C

[a] In the host, resistance is conferred by single dominant genes (R1).

[b] In the pathogen, virulence is conferred by recessive genes (a1). Avirulence is conferred by dominant genes (A1).

[c] Host is resistant. Incompatible interaction.

[d] Host is susceptible. Compatible interaction.

Isolation of the gene does not guarantee that we will know the function of the gene product, but the gene product can be localized and the expression of the gene studied and altered to determine the effects on race-specific resistance.

Legumes have symbiotic as well as pathogenic interactions with microorganisms. Recent experiments have identified some of the plant signals sent to symbionts as intermediates of the flavonoid/isoflavonoid pathway.[19,20] Portions of this pathway are activated to produce phytoalexins in response to pathogens. This suggests that the regulation of this pathway is central to the plant's response to microorganisms.

RACE-SPECIFIC RESISTANCE OF SOYBEAN TO ROOT AND LEAF PATHOGENS

We are studying the interaction of soybean roots with the soilborne fungus, _Phytophthora megasperma_ f.sp. _glycinea_ (Pmg) which causes root rot and the interaction of soybean leaves with the airborne bacterial pathogen Pseudomonas syringae pv. glycinea (Psg) that causes blight. Our method of analyzing the interaction is to look at the time course of induction of specific mRNAs after inoculation of the host with a virulent race of the pathogen (the host is susceptible, the interaction is compatible) or an avirulent race inducing the hypersensitive response (the host is resistant, the interaction is incompatible). We have tried to select pathogen races that are isogenic and have only used one host cultivar to decrease variability in the response due to unknown genetic factors in the host or pathogen.

The disease resistance response requires the transcriptional activation of a number of genes,[21] some of which are involved in phytoalexin biosynthesis. To measure the induction of mRNA of an enzyme involved in phytoalexin production, we have used a chalcone synthase cDNA from Phaseolus vulgaris.[6] To get a qualitative idea of how widespread the activation is, we have looked at mRNA levels for calmodulin, ubiquitin and hydroxyproline-rich glycoprotein. These probes were chosen because they are highly conserved throughout nature and heterologous probing is possible. The calmodulin probe is from the electric eel.[22] Calmodulin is involved in Ca^{++} binding. The ubiquitin probe is from humans.[23] Ubiquitin is involved in protein turnover, especially that of histones. The hydroxyproline-rich glycoprotein (HRGP) probe is from carrot.[24] HRGP is a cell wall protein that is involved in wound repair and the wound response.[25]

ROOT RESPONSE TO PHYTOPHTHORA MEGASPERMA f.sp. GLYCINEA

Phytophthora megasperma f.sp. glycinea (Pmg) is a soilborne Oomycete that causes root rot of soybeans.[26,27] The genus Phytophthora contains many phytopathogenic species; the best known is Phytophthora infestans, the causal agent of potato late blight. Pmg can be a serious problem on soybeans in poorly drained soils and during cold rainy Springs. Flooding in the field induces the production of a motile zoospore that swims through the ground water and attaches on plant roots. Infection of young seedlings may cause death. Older plants with larger root systems will not be killed by the fungus, but are more likely to wilt during dry periods. The fungus can infect the plant at any stage of root development. Pmg is capable of biotrophic growth and does not produce a toxin that kills the cells in advance of the growing hyphae. Plants can survive through the growing season with a Pmg infection. Pmg grows upward in the plant and, late in the growing season, infected plants can be seen with lesions extending up the stem.

The genetics of race-specific resistance to Pmg have been extensively studied.[28] Six race-specific resistance loci have been identified. There are 26 reported Pmg races.[29] Pmg was first isolated in Ohio in 1950. The next Midwestern race was observed in 1973, growing on plants that contained a Pmg resistance gene. Since 1973, 23 races have been reported. Some of the increase in races is due to testing isolates on more soybean cultivars so that finer pathogenic phenotypes can be detected. However, the number of new races reported is strong evidence that Pmg can evolve new races rapidly, which is unusual for a soilborne pathogen. Because Pmg is strictly homothallic (does not outbreed), new races are not created through sexual recombination.

We have studied Pmg genetics to determine how closely related the different Pmg races are and to determine if avirulence is dominant to virulence, as predicted by Flor's hypothesis. We observed no isozyme differences at 19 different loci for Pmg races.[30] Recent studies with Pmg heterokaryons also demonstrate that avirulence is dominant in Pmg;[31] thus the Pmg-root interaction follows Flor's gene-for-gene hypothesis.

We characterized the mRNAs from zoospore inoculated roots as our assay of the race-specific resistance response. Two-day old soybean seedlings resistant to Race 1 and susceptible to Race 3 were individually inoculated by incubation in a solution of about 10,000 zoospores of Race 1 or Race 3 for 2 hours. Inoculated seedlings were transferred to sterile water and 35 seedling roots were harvested at each time point for isolation of RNA. Five control seedlings from each time point were planted in vermiculite and scored for disease symptoms after 2 or 3 days. Control seedlings inoculated with Race 1 grew as well as those mock-inoculated in water but they had a necrotic area on the tip of the root. Secondary roots developed rapidly and were not discolored. Seedlings

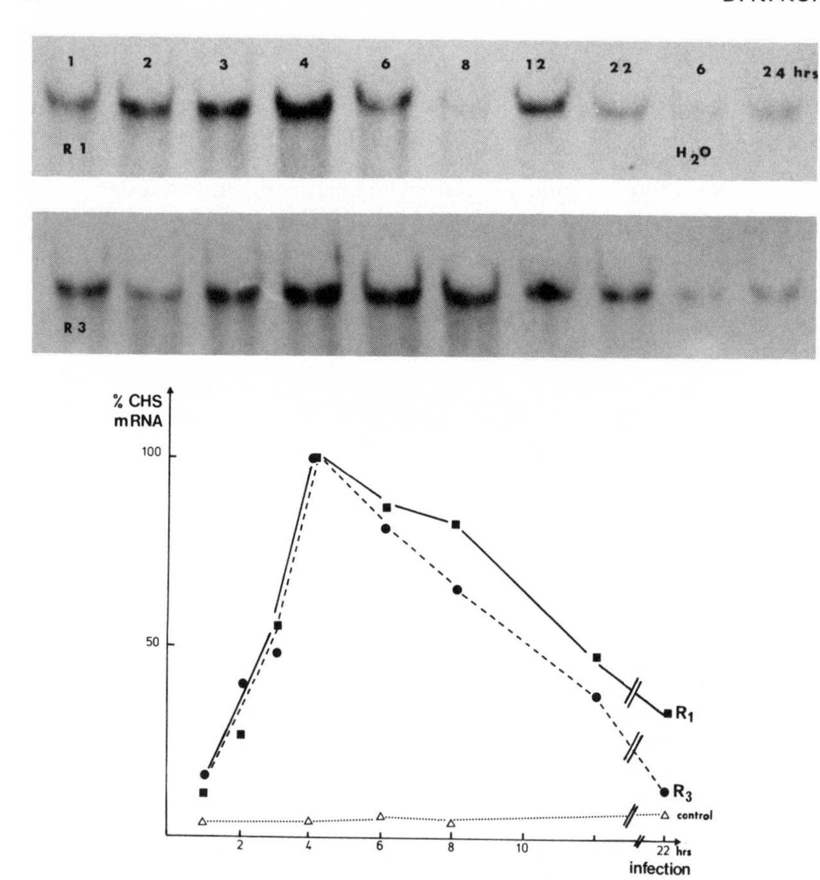

Fig. 2. Autoradiogram of total RNA from soybean cv. Pella roots inoculated with Pmg Race 1 (avirulent) or Pmg Race 3 (virulent). Roots were harvested at the times indicated after inoculation. CHS mRNA was detected by hybridization with a *Phaseolus vulgaris* CHS cDNA. Graph shows scanning densitometry data from a similar experiment.

inoculated with Race 3 were either dead or heavily colonized with mycelium. Growth was stunted and the roots appeared brown and soft. Seedlings had few secondary roots and all secondary roots were infected.

We have seen no difference in the kinetics of induction of chalcone synthase mRNA (Fig. 2) or HRGP mRNA

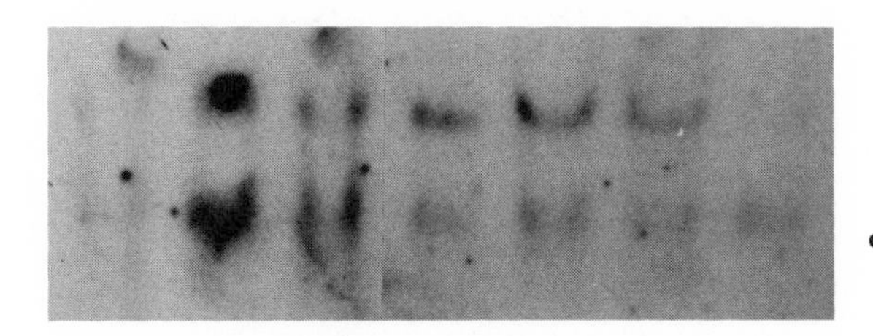

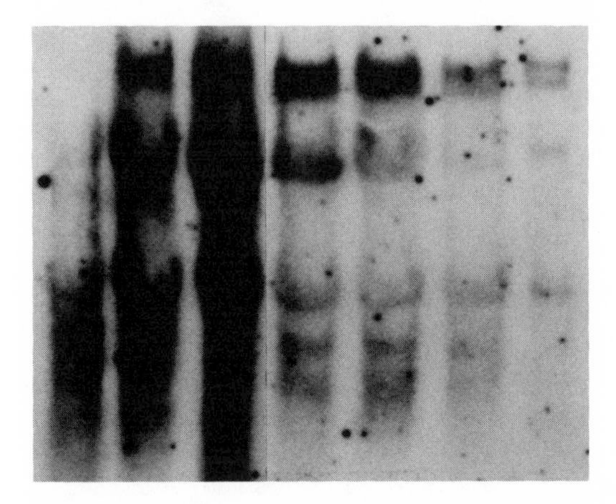

Fig. 3. Autoradiograms of total RNA from soybean roots inoculated with Race 1 (avirulent). Roots were harvested at the times indicated after inoculation. Calmodulin mRNA was detected with an electric eel calmodulin cDNA. Hydroxyproline-rich glycoprotein (HRGP) mRNA was detected with a carrot HRGP gene.

(Figs. 3 and 4) after inoculation of the roots with Race 1 (avirulent) or Race 3 (virulent). There is variability in the amplitude and exact time of the peak of the response from experiment to experiment but no consistent, race-specific differences. In each experiment, CHS and HRGP mRNA are induced in both Race 1 and Race 3 inoculated roots; they reached a maximum between 4-6 hours and return to preinduction levels by 24 hours after pathogen inoculation. Calmodulin mRNA (Figs. 3 and 4) and ubiquitin mRNA (see Fig. 10) are not induced. Inoculation of the roots

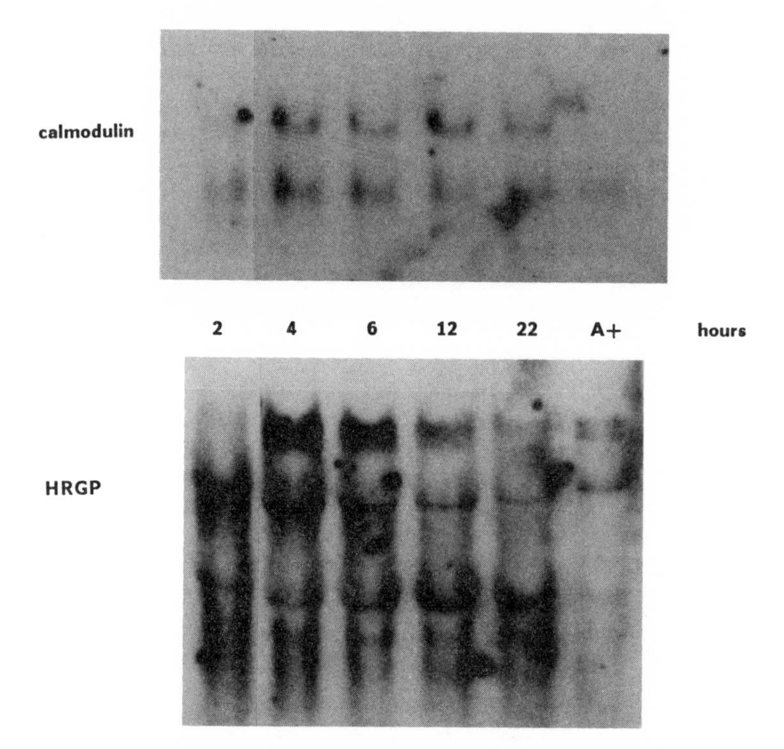

Fig. 4. Autoradiograms of total RNA from seedlings inoculated with R3 (virulent). Roots were harvested at the times indicated after inoculation. Calmodulin and HRGP mRNAs were detected as in Figure 3.

with zoospores of a non-host pathogen, P. megasperma f.sp. medicaginis (Pmm), does not cause a noticeable hypersensitive response or an induction of CHS mRNA (data not shown). Pmm is a pathogen of alfalfa, but may be able to penetrate the epidermis of soybean roots, as Pmg zoospores are able to penetrate the epidermis of alfalfa.[32] Thus, the roots are making a specific recognition of a pathogen, but the response we are measuring (the induction of CHS and HRGP mRNA) is not race-specific. The roots are responding race-specifically as control roots inoculated with zoospores of Race 1 show hypersensitive resistance, and roots inoculated with Race 3 zoospores show no hypersensitive response and are susceptible.

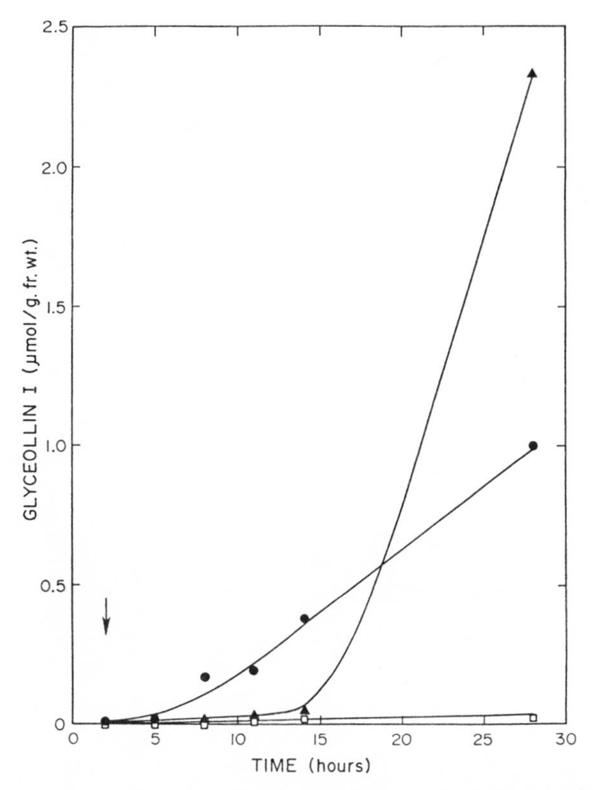

Fig. 5. Quantitation of glyceollin I by radioimmunoassay in single whole roots. Reprinted from Hahn _et al._[33] with permission.

Other groups have studied the interaction of soybean roots with Pmg zoospores with regards to enzyme activity and glyceollin accumulation. Using a radioimmunoassay for glyceollin, Hahn _et al._[33] have measured glyceollin accumulation in zoospore affected roots as a function of time, cellular location and race. There is a 12-hour lag in the overall accumulation of glyceollin in roots inoculated with R3 (virulent) (Fig. 5). Glyceollin accumulation is greater in the sections nearest the epidermis in both the R1 and R3 inoculated roots, with no significant differences in accumulation seen at the earliest time point (5.5 hours). Hahn _et al._[33] and Beagle-Ristaino and Rissler[34] observed that there are no significant differences in the growth of R1

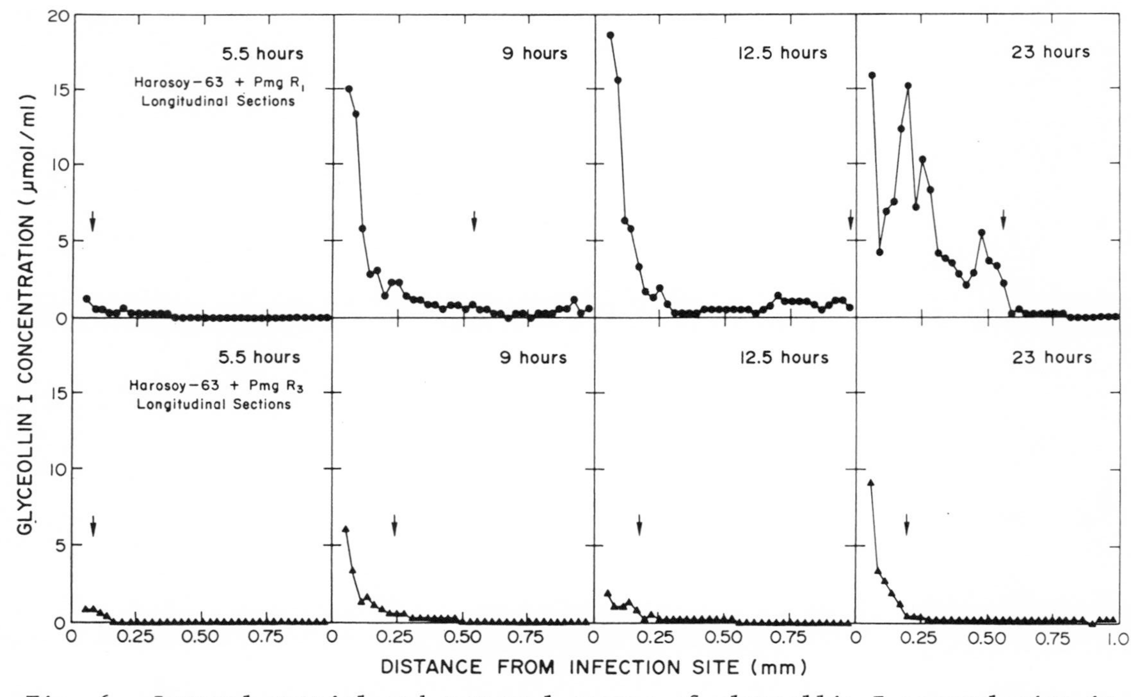

Fig. 6. Lateral spatial and temporal course of glyceollin I accumulation in single soybean roots infected with Race 1 or Race 3 of Pmg. Reprinted from Hahn *et al.*[33] with permission.

and R3 for the first four hours after infection. Beagle-Ristaino and Rissler noted that both races penetrated the epidermis mainly by growing through anticlinal walls. In Rl inoculated roots, a hypersensitive response was visible within 5 hours of inoculation, before significant glyceollin accumulation was detected.[33] In Figure 6, phytoalexin accumulation is observed in cells 0.5 mm distant from the inoculation site and the site of the hypersensitive response. Thus, even in resistant roots expressing the hypersensitive response, phytoalexin accumulation does not necessarily lead to cell necrosis. No hypersensitive response was seen in the susceptible roots even at 23 hours when glyceollin levels at the inoculation site were well above the toxic level for Pmg.

Bonhoff _et al._[35] have measured the activity of several of the enzymes of glyceollin biosynthesis in soybean roots inoculated with Rl and R3 zoospores. Enzyme assays were done on homogenates of 18 roots for every time point. Induction of phenylalanine ammonia-lyase (PAL) and chalcone synthase (CHS) was significantly higher in Rl inoculated roots, but R3 inoculated roots also showed induction of PAL and CHS. There was no significant induction of isoflavone synthase until about 7 hours after inoculation. Dihydroxypterocarpan 6a-hydroxylase, an enzyme much later in the glyceollin biosynthetic pathway (see Fig. 1) showed only a significant increase in activity in the incompatible interaction (after inoculation with Rl), although both R3 incoulated and water controls showed a reproducible increase in activity with time. No significant differences were observed for glucose-6-phosphate dehydrogenase and glutamate dehydrogenase activity between inoculated and control roots.

Hahn _et al._[33] and Bonhoff _et al._[35] both observed an early increase in either glyceollin accumulation or enzyme activity in both Rl and R3 inoculated roots. Glyceollin accumulation was more rapid and reached higher levels, and the enzymes for phytoalexin biosynthesis were induced to greater activity in Rl inoculated roots. However, both Rl and R3 inoculated roots have an early response to infection and, at least at the level of mRNA induction, this early response is identical in Rl and R3 inoculated roots. The only enzyme that showed a clear race-specific induction was dihydroxypterocarpan 6a-hydroxylase.

LEAF RESPONSE TO PSEUDOMONAS SYRINGAE pv. GLYCINEA

P. syringae pv. glycinea (Psg) causes bacterial blight of soybean.[36] Susceptible plants show chlorotic spots and water soaking of infected leaves. Resistant plants respond with the hypersensitive response and only small necrotic flecks are seen on the leaves. Seven races of Psg have been defined by their growth on a set of soybean differential cultivars (different from those used to define Pmg races). Staskawicz et al.[37] have isolated an avirulence gene, avrA, from Psg Race 6. Two other avirulence genes (avrB and avrC) have been isolated from Race 0. Transformation of an avirulence gene into a virulent race will make the race avirulent on a particular soybean cultivar. For example, R4 can infect Harasoy and cause chlorosis. R4 with the avrA gene can not infect Harasoy, which responds with the hypersensitive response. R4 and R4 avrA differ only by a single gene, so that the differential response of the plant to infection is probably due to the avirulence gene product. The avrA gene is constitutively expressed in Psg, unlike the avrB and avrC genes that are not expressed until the bacterium interacts with the plant (Keen, personal communication).

We have characterized the induction of CHS, calmodulin and ubiquitin mRNA in soybean leaves vacuum infiltrated with Psg R4 (compatible interaction) or Psg R4avrA (incompatible interaction). The induction of CHS mRNA is clearly race-specific. Leaves vacuum infiltrated with Psg R4avrA show a strong hypersensitive response within 24 hours. Since almost every cell in the leaf has come into contact with the pathogen, the entire leaf dies. By 24 hours, the leaves are wilted and at 48 hours have become brown and dry. Leaves vacuum infiltrated with R4 show no symptoms in 24 hours but chlorotic spots over all the leaf appear between 48 and 72 hours. CHS mRNA is strongly induced in leaves inoculated with R4avrA and weakly induced in leaves inoculated with R4 (see Figs. 7 and 8). Calmodulin (data not shown) and ubiquitin mRNA levels do not increase in response to inoculation (Fig. 9). The peak of CHS mRNA induction is about 8 hours after inoculation and CHS mRNA levels return to preinoculation levels by 24 hours. No information is available on CHS enzyme activity or glyceollin accumulation in vacuum infiltrated leaves. Vacuum infiltration increases the number of responding cells and the number of cells that die; this makes comparison

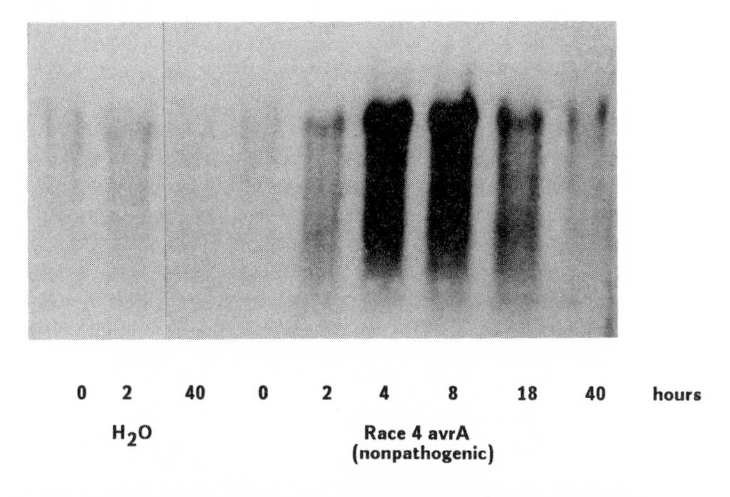

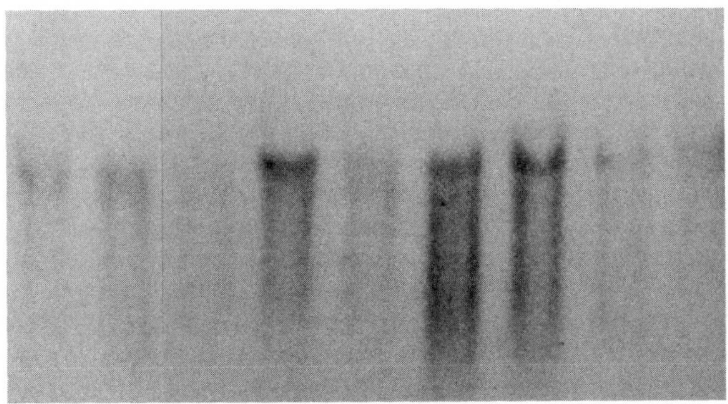

Fig. 7. CHS mRNA in leaves inoculated with nonpathogenic and pathogenic races of Psg. Autoradiogram of total RNA from soybean leaves vacuum infiltrated with Pseudomonas synringae pv. glycinea R4 or R4avrA. CHS mRNA detected as in Figure 2. Leaves were harvested at the times indicated after inoculation.

difficult with other inoculation methods where there is more living tissue to produce phytoalexins. Accumulation of glyceollin is increased in leaves inoculated by another method with an avirulent race.[38] However, glyceollin is also produced in leaves inoculated with a virulent race.

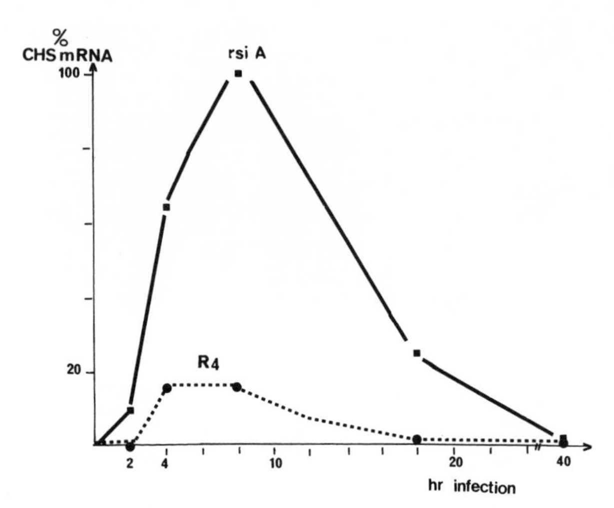

Fig. 8. Scanning densitometry of the autoradiogram in Figure 7. rsiA = R4avrA

COMPARISON OF THE RACE-SPECIFIC RESISTANCE RESPONSE IN ROOTS AND LEAVES

We have seen two very different responses when looking at two different race-specific, resistance interactions of soybean. From probing of polyA+ RNA of uninoculated plants (data not shown), we have seen that CHS and HRGP mRNA are relatively more abundant in the roots than in the leaves. Calmodulin mRNA is relatively more abundant in leaves than roots (data not shown). After inoculation of leaves and roots with pathogen there is no significant increase in calmodulin or ubiquitin mRNA. Thus, we know that the increase in CHS mRNA and HRGP mRNA is not a reflection of a general increase in all mRNA species.

Why is the induction of CHS mRNA in response to pathogen inoculation so different in roots and leaves? The plant response to stress varies from organ to organ. Plant roots are much less likely to respond to heat shock or cold stress because they are in soil and are physically buffered from such shocks. On the other hand, roots are constantly in contact with microorganisms, and survival would be difficult if cell necrosis (hypersensitive

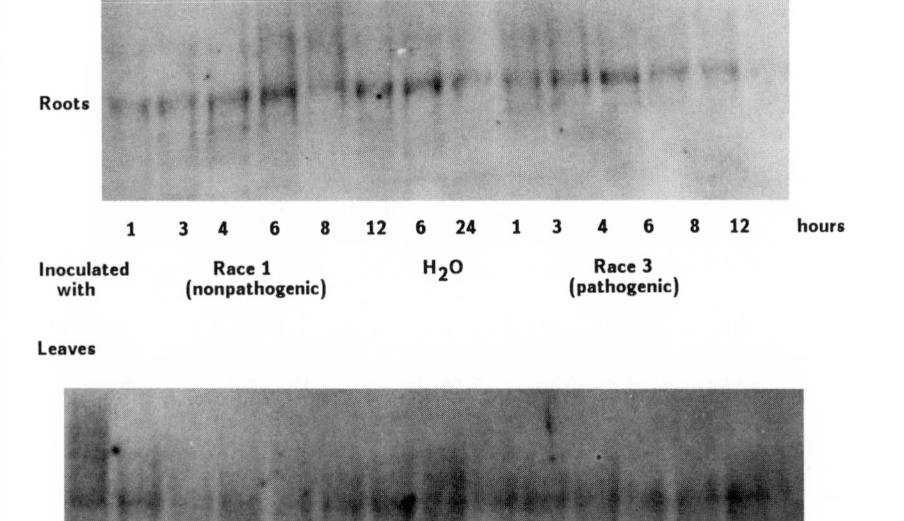

Fig. 9. Ubiquitin mRNA in pathogen inoculated roots and leaves. Autoradiogram of total RNA from roots inoculated with PMG R1 or R3 leaves vacuum infiltrated with Psg R4 or Psg R4avrA. Ubiquitin mRNA detected with a human ubiquitin cDNA. Controls are mock inoculations with water.

response) resulted from every interaction with a microorganism. Water stress affects roots less than the aerial parts of plants, as the water potential may be low in the soil but is much lower in the air. Insects can attack either roots or leaves, but it is doubtful if proteinase inhibitors accumulate in the leaves after wounding of the roots. In all responses to stress, the roots and leaves act almost as two different, distinct organisms (aerial and subterranean) in a very tight symbiosis. Thus, one might expect to see differences in the response of roots and leaves to pathogen inoculation.

We see the identical induction of CHS and HRGP mRNA in R1 and R3 inoculated roots. The response is rapid, peaking within 4-6 hours after inoculation. The zoospores have not had long to penetrate, so the speed of the response suggests that tissues at or near the surface are responsible. Beagle-Ristaino and Rissler[34] reported that zoospore penetration of the epidermis occurs between 0.5 and 1.5 hours after inoculation. Hahn et al.[33] and Bonhoff et al.[35] saw similar increases in glyceollin and glyceollin biosynthetic enzymes until about 6 hours after inoculation. The early induction of CHS and HRGP mRNA may be the general alert response of the root whenever a microorganism penetrates the epidermis. Whatever determines race-specific resistance occurs after this initial stage. Perhaps it is the induction of enzymes late in the glyceollin biosynthetic pathway. Thus, the product of the race-specific resistance gene must act at a stage in the response after the initial recognition of the pathogen.

Since roots interact with numerous microorganisms they must be able to distinguish between pathogens and symbionts. The induction we are seeing may be the early warning, with the plant deciding at the next stage whether the invader should be stopped or encouraged to grow. Recent evidence of plant factors that can encourage or discourage nodulation of Rhizobium[13] show there are roles for flavonoids and isoflavonoids in the plant-microorganism interaction besides that of phytoalexins. The induction we see of CHS may be part of a signal to the microorganism to discourage growth in a certain region of the root.

Another explanation of the induction of CHS mRNA is that differential expression of CHS genes is occurring in the root. In *P. vulgaris*, there are 6 to 8 CHS genes.[39] Different subsets of these genes are induced in cells after pathogen inoculation, elicitor treatment, wounding or illumination. The genes show differential induction in both the timing and amount of transcription in cells. In our experiments, the cDNA probe we used was not capable of discriminating among CHS transcripts from different genes. Thus, there may have been a race-specific induction of a particular CHS gene in the roots that we could not detect against the background of a non-race-specific increase in total CHS mRNA. To study differential induction of CHS genes, we have isolated four putative CHS genes from soybean that differ in their restriction patterns. We are in the process of identifying gene specific regions from these clones that will allow us to quantitate the amount of transcription from each gene. We have qualitative evidence suggesting that CHS mRNA is more abundant in the roots than in the leaves of uninoculated plants. We want to determine which of the CHS genes are expressed constitutively and which are induced in response to pathogen inoculation in the roots and leaves. Because the increase in CHS mRNA is transient, we are also interested in measuring the stability of the CHS mRNA from constitutively expressed and inducible genes in both roots and leaves.

In the leaves, the response to pathogen inoculation is much different than in the roots. First, leaf cells never form symbioses with microorganisms. Second, large amounts of bacteria are vacuum infiltrated into the intercellular space, which simulates a later time point in an event (the growth of the bacteria to a level where the plant would respond) that would normally occur over a period of days. Third, we are clearly involving almost all of the cells in the leaf, as the entire leaf shows the hypersensitive response, becoming necrotic, withering and dying, and drying up. Therefore, we have also strongly synchronized the response; this is not really possible in the roots, where attachment and growth of zoospores occurs over a period of two hours and only the outer surface of the root is in contact with the pathogen.

In the leaves, race-specific resistance could be triggered at the initial stage of recognition of the pathogen. In resistance, we see a large increase in CHS mRNA and no increase in calmodulin or ubiquitin mRNA. In susceptibility, we see no significant increase in CHS mRNA or in ubiquitin or calmodulin mRNA. It is almost as if there were no recognition of the pathogen by the plant. The product of a leaf race-specific resistance gene could be a receptor that triggers the hypersensitive response.

We tried to examine the differences in response of leaves and roots by vacuum infiltrating the leaves with Pmg zoospores and inoculating roots with Psg. Neither infection of susceptible plants nor hypersensitive response in the roots or leaves was observed. We have recently been able to infect susceptible leaves with Pmg zoospores by dipping the leaves in a zoospore solution and then incubating the plants under high humidity. We will test to see if CHS mRNA is induced only in the resistant case. Such an induction would be excellent evidence that the race-specific resistance response is modified to meet the needs of the organ. Race-specific resistance has been studied in soybean hypocotyls and cotyledons, and these organs do express race-specific resistance to Pmg. However, biochemical studies are difficult to interpret because the inoculation of hypocotyls and cotyledons is usually by the introduction of Pmg mycelium into a wound.[40,41]

Is the triggering of the race-specific resistance response in leaves as simple as it appears? Keen and Staskawicz (personal communication) have found that the hypersensitive response in leaves can differ in relation to the avirulence genes present in the pathogen. Race 0 of Psg contains avrB and avrC and gives a normal, rapid hypersensitive response that develops in about 24 hours. When these genes are put separately into a virulent race, they cause the bacteria to be avirulent, but the response of the plant is not the same hypersensitive response as seen to Race 0. AvrB incites the hypersensitive response in about 16 hours while avrC does so in 36 to 48 hours. The avrC gene product has been isolated after production in Escherichia coli. When this protein is inoculated onto leaves alone, there is no response. When this protein and live virulent bacteria are inoculated on leaves, there is no hypersensitive response. This suggests that the timing or location of expression of the avirulence gene may be important in triggering the hypersensitive response. It does not support the presence of a cell surface receptor on leaf cells that mediates race-specific resistance or a simple recognition event as the trigger for the hypersensitive response.

Thus, the data are not easily explained by the simple models for race-specific resistance. In the roots, the early steps of phytoalexin biosynthesis are equally induced in response to a virulent or avirulent race. This would suggest that there is no involvement of race-specificity as the first stage of recognition. In the leaves where a simple receptor model fits the CHS induction data, the data on responses to avrB and avrC suggest that more than one signal is needed to induce the hypersensitive response. What are the current models of race-specific resistance and can they explain such data?

CURRENT MODELS FOR RACE-SPECIFIC RESISTANCE

The simplest model for a response is the single hit receptor model, where a single event leads to a response. Initial models should always assume the simplest case, until enough data is collected to suggest that the model is no longer useful as a means to think about the problem. Perhaps we are now at this stage with regard to plant-microbe interactions and, particularly, to the disease resistance response.[42]

There are two closely related models to explain the biochemical basis of race-specific resistance.[43-45] The dimer model postulates a direct interaction between the host resistance gene product and the pathogen avirulence gene product. This interaction triggers the hypersensitive response. The host resistance gene product is a cell surface receptor and the pathogen avirulence product is a cell surface protein. The second model does not require a direct protein-protein interaction between the host and pathogen gene products. The interaction can be between the product of the reaction catalyzed by the avirulence gene product and the host resistance gene product. This pathogen product may itself act as a signal or the host resistance gene product may act on this signal to produce another signal. For example, the pathogen gene product might be a glycosyl transferase affecting the pathogen cell wall structure or the glycosylation pattern on an extracellular glycoprotein. The host resistance gene product might be a glycosidase that could recognize this particular glycosyl linkage in the cell wall or extracellular glycoprotein and clip out a particular glucan fragment. This fragment could then interact with some type of receptor in the host cell to trigger the race-specific resistance response.

A serious question about this second model is whether such a mechanism can explain the genetic data (Ellingboe, personal communication). Since resistance which is not triggered by a direct interaction between the avirulence gene product and the resistance gene product could be triggered by other mechanisms, race-specific resistance would not map to a single locus in the host. It could be possible, for example, that another host glucosidase with specificity similar to the resistance gene product could also trigger resistance to a particular pathogen; resistance could also be a result of a change in the receptor protein that altered its specificity.

Aside from these technical points, the models are both simple linear models that require only a single initiating event to lead to race-specific resistance. Both models predict the presence of race-specific elicitors, isolatable race-specific factors able to trigger the hypersensitive response and host receptors that are specific for these elicitors.

Elicitors have been studied intensively, particularly the glucan elicitors from the cell walls of Pmg.[47] Pmg elicitors incite responses in a wide variety of plants, while Pmg itself infects only soybean. There is no convincing evidence that race-specific elicitors can be isolated from Pmg cell walls.[47] Further, the failure of the product of a Psg avirulence gene to incite a race-specific response strongly suggests that models based on a single event activation of the response are inadequate.

MULTIPLE INTERACTIONS ARE NECESSARY FOR RACE-SPECIFIC RESISTANCE

Perhaps the linear models can be improved by viewing the interaction as a sequential series of steps, with race-specificity not as one of the initial steps (see Fig. 10). In a sequence, any single step can give specificity, but for race-specific resistance, the specificity factor must act as the last step before the signal to activate the hypersensitive response. Otherwise, a non-specific elicitor or other signals could short circuit the race-specific resistance step. If race-specificity acted at an earlier step (A -- B) than the elicitor, the elicitor could then cause the resistance response even if the race-specific gene product were not present. Since elicitors can be made from either virulent or avirulent races of Pmg, we are reasonably sure that whatever step the cell wall elicitors activate should be before race-specificity.

The linear multistep model suggests that the cell may need more than one signal to respond with a hypersensitive

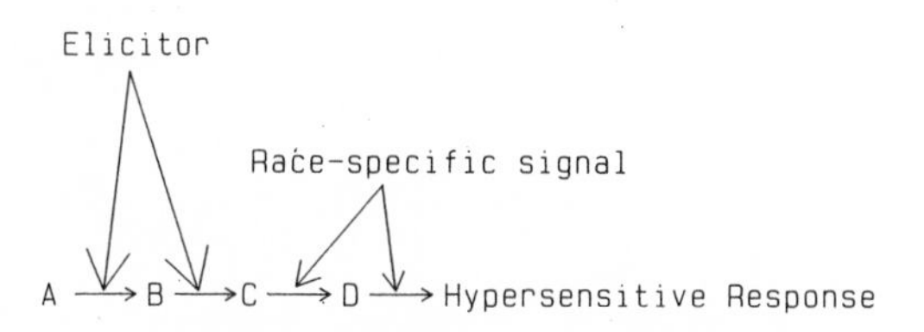

Fig. 10. Linear multistep model. Letters and numbers represent pathway intermediates. Intermediates are not necessarily chemical compounds, they can also represent a charged membrane or a developmental stage.

response. These signals are sequential and thus, inhibition of an earlier step should block the entire sequence. Such a model is sufficient to explain our data on the induction of CHS and HRGP mRNA in soybean roots and inhibition of transcription and translation is known to inhibit the race-specific resistance response.[48] However, data on the inhibition of other induced responses is not as easy to interpret with this model.

One of the earliest responses of a plant to pathogen attack and of plant cell cultures to elicitor treatment[49] is the evolution of ethylene. This response does not require de novo synthesis of enzymes. Ethylene production can be specifically inhibited and the inhibition can be specifically reversed. In the linear multistep model, inhibition of ethylene production should inhibit the rest of the response because ethylene synthesis occurs so early. Similarly, treatment of cells with ethylene should mimic the action of an elicitor. Data on the response of wounded soybean hypocotyls[50] and parsley tissue culture cells[49] to elicitor suggest that neither statement is wholly true. Complete inhibition of ethylene production gives only a partial inhibition of the induction of phytoalexin biosynthetic enzymes.[49,50] Treatment of the tissue with ethylene alone does not induce phytoalexin biosynthesis,[49] although ethylene or elicitor will induce HRGP synthesis in cucumber.[10] Thus, ethylene has an effect but may not be part of a linear sequence of steps leading to resistance.

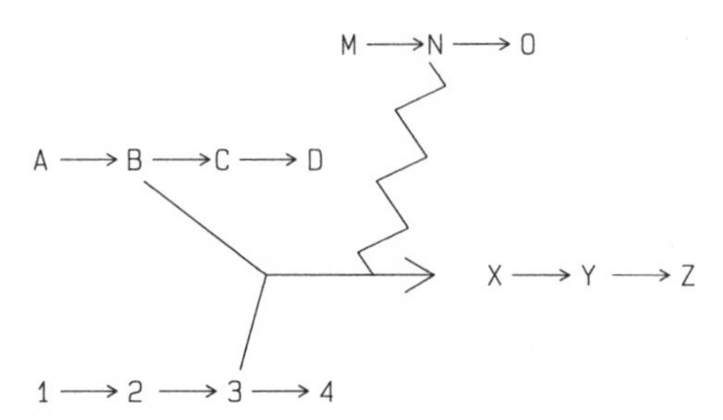

Fig. 11. Parallel multistep model. Letters and numbers represent pathway intermediates as in Figure 10.

PARALLEL MULTISTEP INDUCTION MODEL OF RACE-SPECIFIC RESISTANCE

A model that explains the ethylene data involves several parallel multistep sequences that must act in concert to trigger the resistance response. Such a complex model also has advantages in explaining environmental effects on race-specific resistance and field tolerance of susceptible cultivars. Finally, the parallel multistep model suggests a possible mechanism for the plant to modulate and turn off the response.

The model assumes that at least two separate sequential series of events (1,2,3,4 and A,B,C) are necessary to induce the hypersensitive response. The timing of induction and the location of the events is also important. In Figure 11, two intermediates (B and 3) interact to induce X,Y and Z. The hypersensitive response is not necessarily X,Y and Z alone, but rather the synergistic interaction of all the pathways. To achieve this synergistic interaction, 1,2,3,4 and A,B,C must be initiated in the same cell at the same time. Delay or inhibition of one of the steps will then alter the response of the plant by reducing the synergism. Activation of one of the pathways alone, even X,Y,Z will not lead to the hypersensitive response. The race-specific signal can still act as a trigger and it can act as such in any one of the multiple pathways involved in the response.

We have avoided discussing environmental effects on race-specific resistance because such factors are controlled in the laboratory. In the field, plants respond to a variety of environmental factors. The physiological and environmental state of the plant affects and, in some cases, can reverse race-specific resistance.[50-52] Also, cultivars that show no race-specific resistance to infection under laboratory conditions can show resistance in the field.[20] This type of resistance (field tolerance) is not absolute like race-specific resistance. The plants can be infected but the yield loss due to infection is less. Such field tolerance could be explained by invoking the multistep parallel event model, where a pathway initiated by environmental factors (M, N, O) leads to a state where the synergism resulting in resistance can take place (see above model).

The synergism resulting from N may not be the hypersensitive response, as N may not work as well as B in inducing X,Y,Z. The hypersensitive response may be an extreme on a continuous spectrum of plant responses. N may be a weaker synergistic factor than B, may not occur in the cell at a high enough concentration, or may be temporally displaced enough to give less than absolute resistance, but may still provide a level of protection better than in the susceptible cultivars. This level of resistance will depend on the interaction of the same pathways involved in the hypersensitive response.

In the parallel events model, multiple factors are involved in the race-specific response. The whole response is not triggered by a single event even though the whole response is under the regulation of a single gene. This situation is analogous to the formation of tumors in animals due to the expression of an oncogene. In a recent experiment,[53] transgenic mice were produced that expressed two oncogenes. Either of these oncogenes alone is enough to cause a tumor and to cause transformation of tissue culture cells. Mice expressing both genes did show an increase in the number of tumors; however, tumors arose on a stochastic basis in only a few tissues. In a particular tissue, all cells expressed the genes at the same levels but not all cells became transformed. Thus, cell, tissue, organ and environmental factors other than the expression of an oncogene are involved in trigerring tumor formation.

It is not possible at this point to distinguish between the parallel and linear multistep model. However, for both models, we can conclude that (1) multiple signals are required to get the complete response and (2) the race-specific factor can be relatively late in the response rather than being the initial receptor.

In both models, an important step is the activation of transcription of specific genes. However, the eventual outcome of the interaction (susceptibility or resistance) may not be a simple function of a single burst of gene activation, but may depend on the timing of the transcription of particular genes, the cells in which transcription is activated, as well as other physiological factors such as ethylene biosynthesis and cell wall repair. Since in inducing the hypersensitive response the correct series of events can lead to the suicide of the cell, it also may be necessary for the plant to regulate the response at a number of post-transcriptional levels to prevent unnecessary necrosis.

Transcription of stress specific genes has been studied in detail in two systems.[6,54,55] In both the Pmg elicitor-parsley cell system and the C. lindemuthianum - P. vulgaris interaction, there is some evidence for post-transcriptional modification. In elicitor treated parsley cells, PAL transcription and PAL mRNA levels increase to and remain at a maximum while the translational activity of the PAL mRNA and PAL enzyme activity transiently increase and then return to preinduction levels.[4] In the C. lindemuthianum - P. vulgaris interaction there are several reports[56,57] of activation at the enzyme level for PAL and chalcone isomerase (CHI). In a recent report on the coordinate induction of PAL, CHS and CHI in P. vulgaris cells treated with elicitor, enzyme activity continued to increase linearly for at least 15 hours after the induced mRNA levels had returned to near pre-induction levels.[55] These data suggest that there is post-transcriptional control of key regulatory enzymes in phytoalexin production.

The parallel multistep model will only be useful if it stimulates new ways of thinking about plant-microbe interactions. What approaches to understanding plant microbe interactions does the model suggest?

1. One is to broaden the assays used to quantitate the response. It is important to assay a number of factors at the molecular as well as higher levels. In race-specific response studies, the timing of appearance of the hypersensitive response can be used as an assay of the response of the plant at the tissue level, while phytoalexin synthesis and ethylene evolution are measured at the cellular level.

2. Analyze the kinetics of the various responses for potential synergism. The model suggests that it is not what happens first that will be most important, but what events occur simultaneously in the cell. The synergism of the various pathways is more important than the order in which they are induced.

3. Localize the responses. For a synergistic effect, the various pathways must occur at the same time and in the same place. In race-specific resistance, the major response is localized to the site of infection. Not every cell in the tissue is responding to the pathogen. In quantitating phytoalexins or enzyme activity or mRNA amounts, mixing of responding and nonresponding cells is unavoidable. Molecular probes such as antibodies for stress-induced proteins or cDNAs for stress-induced mRNA should be used to identify the tissues and the cells that are actually responding to the stress.

4. Broaden the study of environmental and physiological effects. In soybeans, phytoalexin synthesis is one aspect of the response to pathogen stress. Phenylalanine is the precursor of the soybean phytoalexin. Kimpel and Kosuge[50] have reported that decreasing the availability of phenylalanine by removing the cotyledons of soybean seedlings reduces both phytoalexin synthesis and the synthesis of PAL in hypocotyls that have been injected with elicitor. In elicitor treated hypocotyls incubated in the dark for 24 hours, there was reduction of phytoalexin accumulation by 50% when compared with plants in the light. What would be the effect of cotyledon removal on the race-specific resistance in the leaves or in the roots? How would inoculation of plants followed by dark incubation affect resistance? Would the hypersensitive response be altered or delayed?

The discussion of the models for the disease resistance response and the suggestions for further research at all levels of the response are not limited to race-specific resistance. I hope they will stimulate thought about other plant responses and lead to new hypotheses about plant responses as well as to the design of the experiments to test such hypotheses.

ACKNOWLEDGMENTS

I would like to thank C.J. Lamb for the CHS cDNA, Anthony Means for the calmodulin cDNA, Jens Vuust for the ubiquitin cDNA and Joe Varner for the HRGP gene. I also thank A.H. Ellingboe and Noel Keen for discussions relating to race-specific resistance.

REFERENCES

1. DIXON, R.A., P.M. DEY, C.J. LAMB. 1983. Phytoalexins: enzymology and molecular biology. In Advances in Enzymology, Related Areas in Molecular Biology. (A. Meister, ed.), Vol. 55, John Wiley and Sons, Inc., New York, 1-136.
2. BAILEY, J.A., J.W. MANSFIELD, eds. 1982. In Phytoalexins. John Wiley and Sons, Inc., New York, 334 pp.
3. HAGMANN, M.-L., H. GRISEBACH. 1984. Enzymatic rearrangement of flavanone to isoflavone. FEBS Lett. 175: 199-202.
4. HAGMANN, M.-L., W. HELLER, H. GRISEBACH. 1984. Induction of phytoalexin synthesis in soybean. Sterospecific 3,9-dihydroxypterocarpan 6a-hydroxylase from elicitor-induced soybean cell cultures. Eur. J. Biochem. 142: 127-131.
5. ZAHRINGER, U., J. EBEL, L.J. MULHEIRN, R.L. LYNE, H. GRISEBACH. 1979. Induction of phytoalexin synthesis in soybean: dimethylalylpyrophosphate: trihydroxypterocarpan dimethylallyl transferase from elicitor-induced cotyledons. FEBS Lett. 101: 90-92.
6. CHAPPELL, J., K. HAHLBROCK. 1984. Transcription of plant defense genes in response to UV light or fungal elicitor. Nature 311: 76-78.

7. EDWARDS, K., C.L. CRAMER, G.P. BOLWELL, R.A. DIXON, W. SCHUCH, C.J. LAMB. 1985. Rapid transient induction of phenylalanine ammonia-lyase mRNA in elicitor treated bean cells. Proc. Natl. Acad. Sci. USA 82: 6731-6735.
8. KREUZALER, F., H. RAGG, E. FAUTZ, D.N. KUHN, K. HAHLBROCK. 1983. UV induction of chalcone synthase mRNA in cell suspension cultures of Petroselinum hortense. Proc. Natl. Acad. Sci. USA 80: 2591-2593.
9. RYDER, T.B., C.L. CRAMER, J.N. BELL, M.P. ROBBINS, R.A. DIXON, C.J. LAMB. 1984. Elicitor rapidly induces chalcone synthase mRNA in Phaseolus vulgaris cells at the onset of the phytoalexin defense response. Proc. Natl. Acad. Sci. USA 81: 5724-5728.
10. ROBY, D., A. TOPPAN, M.T. ESQUERRE-TUGAYE. 1985. Cell surfaces in plant-microorganism interactions: 5. Elicitors of fungal and of plant origin trigger the synthesis of ethylene and of cell wall hydroxyproline-rich glycoprotein in plants. Plant Physiol. 77: 700-704.
11. KEEN, N.T., J.D. PAXTON. 1975. Coordinate production of hydroxyphaseollin and the yellow-fluorescent compound PAK in soybeans resistant to Phytophthora megasperma var. sojae. Phytopathology 65: 635-637.
12. BRUEGGER, B.B., N.T. KEEN. 1979. Specific elicitors of glyceollin accumulation in the Pseudomonas glycinea-soybean host-parasite system. Physiol. Plant Pathol. 15: 69-78.
13. SHOWALTER, A.M., J.N. BELL, C.L. CRAMER, J.A. BAILEY, J.E. VARNER, C.J. LAMB. 1985. Accumulation of hydroxyproline-rich glycoprotein mRNAs in response to fungal elicitor and infection. Proc. Natl. Acad. Sci. USA 82: 6551-6555.
14. TOMIYAMA, K. 1982. Hypersensitive cell death: its significance and cell physiology. In Plant Infection: The Physiological and Biochemical Basis. (Y. Asada et al., eds.), Japan Scientific Society Press, Tokyo/Springer-Verlag, Berlin, pp. 329-344.
15. KUHN, D.N. 1987. Plant responses to stresses at the molecular level. In Plant-Microbe Interactions. (T. Kosuge, E.W. Nester, eds.), Vol. 2, Macmillan Publishing Company, New York, pp. 415-441.
16. GRILL, E., E.-L. WINNACKER, M.H. ZENK. 1985. Phytochelatins: the principal heavy-metal

complexing peptides of higher plants. Science 230: 674-676.

17. RYAN, C.A. 1984. Systematic responses to wounding. In T. Kosuge, E.W. Nester, eds., op. cit. Reference 15, Vol. 1, pp. 307-321.
18. FLOR, A.H. 1947. Host-parasite interactions in flax rust - its genetics and other implications. Phytopathology 45: 680-685.
19. PETERS, N.K., J.W. FROST, S.R. LONG. 1986. A plant flavone, luteolin, induces expression of Rhizobium meliloti nodulation genes. Science 233: 977-980.
20. DJORDJEVIC, M.A., J.W. REDMOND, M. BATLEY, B.G. ROLFE. 1987. Clovers secrete specific phenolic compounds which either stimulate or repress nod gene expression in Rhizobium trifolii. EMBO J. 6: 1173-1179.
21. CRAMER, C.L., T.B. RYDER, J.N. BELL, C.J. LAMB. 1985. Rapid switching of plant gene expression induced by fungal elicitor. Science 227: 1240-1243.
22. LAGACE, L., T. CHANDRA, S.L.C. WOO, R.A. MEANS. 1983. Identification of multiple species of calmodulin messenger RNA using a full length complementary DNA. J. Biol. Chem. 258: 1684-1688.
23. WIBORG, O., M.S. PEDERSEN, A. WIND, L.E. BERGLUND, K.A. MARCKER, J. VUUST. 1985. The human ubiquitin multigene family: some genes contain multiple directly repeated ubiquitin coding sequences. EMBO J. 4: 755-759.
24. CHEN, J., J.E. VARNER. 1985. Isolation and characterization of cDNA clones for carrot extensin and a proline-rich 33-kDa protein. Proc. Natl. Acad. Sci. USA 82: 4399-4403.
25. WILSON, L.G., J.C. FRY. 1986. Extensin - a major cell wall glycoprotein. Plant Cell Environ. 9: 239-260.
26. PAXTON, J.D. 1983. Phytophthora root and stem rot of soybean: a case study. In Biochemical Plant Pathology. (J.S. Callow, ed.), Wiley and Sons, Inc., New York, pp. 19-29.
27. SCHMITTHENNER, A.F. 1985. Problems and progress in controlling Phytophthora root rot of soybean. Plant Dis. 69: 362-368.
28. ATHOW, K.L., F.A. LAVIOLETTE. 1984. Breeding soybeans for race-specific Phytophthora resistance. Proceedings 13th Soybean Seed Research Conference, American Seed Trade Assoc., pp. 12-21.

29. LAYTON, A.C., K.L. ATHOW, F.A. LAVIOLETTE. 1986. A new physiological race of Phytophthora megasperma f.sp. glycinea. Plant Dis. 70: 500-501.
30. LAYTON, A.C., D.N. KUHN. 1988. Heterokaryon formation by protoplast fusion of drug resistant mutants in Phytophthora megasperma f.sp. glycinea. Experimental Mycology (in press).
31. LAYTON, A.C., D.N. KUHN. 1988. The virulence of interracial heterokaryons of Phytophthora megasperma f.sp. glycinea. Phytopathology (in press).
32. MILLER, S.A., D.P. MAXWELL. 1982. Light microscope observations of susceptible, host resistant and nonhost resistant interactions of alfalfa with Phytophthora megasperma. Can. J. Bot. 62: 109-116.
33. HAHN, M., A. BOHNERT, H. GRISEBACH. 1985. Quantitative localization of the phytoalexin glyceollin I in relation to fungal hyphae in soybean roots infected with Phytophthora megasperma f.sp. glycinea. Plant Physiol. 77: 591-601.
34. BEAGLE-RISTAINO, J.E., J.F. RISSLER. 1983. Histopathology of susceptible and resistant soybean roots inoculated with zoospores of Phytophthora megasperma f.sp. glycinea. Phytopathology 73: 590-595.
35. BONHOFF, A., R. LOYAL, J. EBEL, H. GRISEBACH. 1986. Race: cultivar-specific induction of enzymes related to phytoalexin biosynthesis in soybean roots following infection with Phytophthora megasperma f.sp. glycinea. Arch. Biochem. Biophys. 246: 149-154.
36. FAHY, P.C., A.B. LLOYD. 1983. Pseudomonas: the fluorescent pseudomonads. In Plant Bacterial Disease. (P.C. Fahy, G.J. Persley, eds.), Academic Press, New York, pp. 141-188.
37. STASKAWICZ, B.J., D. DAHLBECK, N. KEEN. 1984. Cloned avirulence gene of Pseudomonas syringae pv. glycinea determines race-specific incompatibility on Glycine max. Proc. Natl. Acad. Sci. USA 81: 6024-6028.
38. KEEN, N.T., B.W. KENNEDY. 1974. Hydroxyphaseollin and related isoflavonoids in the hypersensitive reaction of soybeans to Pseudomonas glycinea. Physiol. Plant Pathol. 4: 173-185.

39. RYDER, T.B., S.A. HEDRICK, J.N. BELL, X. LIANG, S.D. CLOUSE, C.J. LAMB. 1987. Organization and differential activation of a gene family encoding the plant defense enzyme chalcone synthase in Phaseolus vulgaris. Mol. Gen. Genet. (in press).
40. BORNER, H., H. GRISEBACH. 1982. Enzyme induction in soybean infected by Phytophthora megasperma f.sp. glycinea. Arch. Biochem. Biophys. 217: 65-71.
41. SCHMELZER, E., H. BORNER, H. GRISEBACH, J. EBEL, K. HAHLBROCK. 1984. Phytoalexin synthesis in soybean (Glycine max). Similar time courses of mRNA induction in hypocotyls infected with a fungal pathogen and in cell cultures treated with fungal elicitor. FEBS Lett. 172: 59-63.
42. RALTON, J.E., M.G. SMART, A.E. CLARKE. 1987. Recognition and infection processes in plant pathogen interactions. In T. Kosuge, E.W. Nester, eds., op. cit. Reference 15, Vol. 2, pp. 217-252.
43. ELLINGBOE, A.H. 1982. Genetical aspects of active defense. In Active Defense Mechanisms in Plants. (R.K.S. Wood, ed.), Plenum Press, New York, pp. 179-192.
44. YOSHIKAWA, M. 1983. Macromolecules, recognition, and the triggering of resistance. In J.A. Callow, ed., op. cit. Reference 26, pp. 267-298.
45. VANDERPLANK, J.E. 1982. Host-pathogen interactions in plant disease. Chapter 6, The gene for gene hypothesis. Academic Press, Inc., New York.
46. KEEN, N.T. 1982. Phytoalexins-progress in regulation of their accumulation in gene-for-gene interactions. In Y. Asad et al., eds., op. cit. Reference 14, pp. 281-299.
47. DARVILL, A.G., P. ALBERSHEIM. 1984. Phytoalexins and their elicitors - a defense against microbial infection in plants. Annu. Rev. Plant Physiol. 35: 243-275.
48. KEEN, N.T., M. YOSHIKAWA. Physiology of disease and nature of resistance to Phytophthora. In Phytophthora, its biology, taxonomy, ecology and pathology. (D.C. Erwin, S. Bartnicki-Garcia, P.H. Tsao, eds.), American Phytopathological Society, St. Paul, Minnesota, pp. 289-301.
49. CHAPPELL, J., K. HAHLBROCK, T. BOLLER. 1984. Rapid induction of ethylene biosynthesis in cultured parsley cells by fungal elicitor and its relation-

ship to the induction of phenylalanine ammonia-lyase. Planta 161: 475-480.

50. KIMPEL, J.A., T. KOSUGE. 1985. Metabolic regulation during glyceollin biosynthesis in green soybean hypocotyls. Plant Physiol. 77: 1-7.
51. MURCH, R.S., J.D. PAXTON. 1980. Rhizosphere salinity and phytoalexin accumulation in soybean. Plant Soil 54: 163-167.
52. CHAMBERLAIN, D.W. 1972. Heat-induced susceptibility to nonpathogens and cross-protection against *Phytophthora megasperma* var. *sojae* in soybean. Phytopathology 62: 645-646.
53. SINN, E., W. MULLER, P. PATTENGALE, I. TEPLER, R. WALLACE, P. LEDER. 1987. Coexpression of MMTV/v-Ha-ras and MMTV/C-myc genes in transgenic mice: synergistic action of oncogenes *in vivo*. Cell 49: 477-485.
54. SOMSSICH, I.E., E. SCHMELZER, J. BOLLMANN, K. HAHLBROCK. 1986. Rapid activation by fungal elicitor of genes encoding "pathogenesis-related" proteins in cultured parsley cells. Proc. Natl. Acad. Sci. USA 83: 2427-2430.
55. LAWTON, M.A., C.J. LAMB. 1987. Transcriptional activation of plant defense genes by fungal elicitor, wounding and infection. Mol. Cell. Biol. 7: 335-341.
56. DIXON, R.A., C. GERRISH, C.J. LAMB, M.P. ROBBINS. 1983. Elicitor-mediated induction of chalcone isomerase in *Phaseolus vulgaris* cell suspension cultures. Planta 159: 561-569.
57. LAWTON, M.A., R.A. DIXON, C.J. LAMB. 1980. Elicitor modulation of the turnover of L-phenylalanine ammonia-lyase in french bean cell suspension cultures. Eur. J. Biochem. 129: 593-601.

Chapter Eight

OLIGOSACCHARIDE SIGNALLING FOR PROTEINASE INHIBITOR GENES IN PLANT LEAVES

CLARENCE A. RYAN

Institute of Biological Chemistry
Washington State University
Pullman, Washington 99164

INTRODUCTION

Plants have evolved a variety of strategies to defend themselves against the multitude of pests and predators that can infect or ingest them.[1] The various defenses that plants employ, often found in combinations, range from physical barriers, such as the cuticle,[2] to a host of toxic or antinutritional chemicals that can be either constitutively synthesized to effective levels or can be induced to accumulate in response to attacking pests.[3] The diversity and complexity of induced plant defensive systems has made it difficult to obtain clear evidence that specific chemicals are responsible for defending the plants against specific intruders. It is only within the past few years that such evidence has begun to emerge, and only a few constitutive or inducible defensive systems have been extensively studied at the biochemical or molecular biological levels. Evidence from studies of inducible plant defenses now strongly implicates a possible role for oligosaccharide fragments derived from fungal and plant cell walls in signalling localized and systemic defensive responses.[4,5,6] Beginning with experiments in the laboratory of Albersheim in the mid-1970s, in which β-glucan components of fungal cell walls were shown to be powerful elicitors of phytoalexin antibiotic synthesis in soybean cotyledons,[7] a concept has emerged from studies in several laboratories that oligosaccharide fragments from fungi and plant cell walls, as well as from insect cuticles, can induce defensive systems of plants when released from sites of pest attacks.

A role for plant cell wall fragments in inducing synthesis of the antibiotic phytoalexins, near sites of fungal infections, resulted from research in the laboratories of West[8] and Albersheim,[9] while a role for pectic fragments in the systemic induction of proteinase inhibitors in plant leaves originated in our own laboratory.[10,11]

Having established that pectic fragments were capable of inducing various defensive responses in plants, these three laboratories, and others, began a search for the specific structural entities within the complex plant cell wall carbohydrates that were responsible for the induction of the responses. In many cases the inducing activity was found to reside in the α-1,4 polygalacturonic acid backbone of the pectic polysaccharides. Localized phytoalexin synthesis in soybean[12] and castor bean[13] cotyledons was elicited by galacturonans (α-1,4 galacturonic acid polymers) of lengths of about 10-14 galacturonic units. Degrees of polymerization (DP) lower than 9 were inactive. On the other hand, oligomers of DP of 2 and larger induced the systemic induction of proteinase inhibitors when supplied to young tomato plants through cut petioles.[14] A comparison of the activities with the sizes of oligomers that elicit glyceollin in soybeans, casbene synthetase in castor beans and proteinase inhibitors in tomato leaves is shown in Figure 1.

Following these initial reports, many defensive responses have now been described that are induced or elicited by plant cell wall fragments or by enzymes that produce these fragments. Table 1 summarizes some of these responses. They include the induction or elicitation of

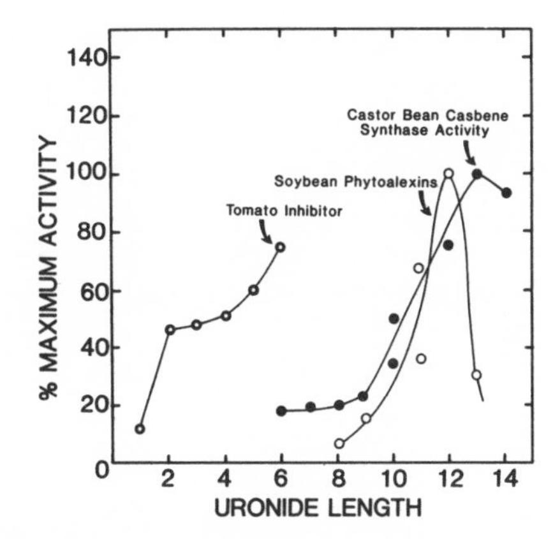

Fig. 1. Relationships of degree of polymerization of α-1,4 galacturonic acid oligomers with proteinase inhibitor inducing activity in tomato leaves (**0**); phytoalexin elicitor activity in cotyledons of soybeans (0); and castor beans (●).

phytoalexins, proteinase inhibitors, tree resins, lignin, ethylene, chitinase, β-glucanase, necrosis and resistance to fungi. This broad diversity of defensive responses to pectic fragments in a wide spectrum of plant genera suggests that intercellular and intracellular mechanism involving these fragments may be a universal property of plants to alert defensive responses.[6]

We have been studying the mechanism of wound-induction of proteinase inhibitors in potato, tomato and alfalfa leaves with the goal of understanding the biochemistry and molecular biology of the communication system involved with oligosaccharide signalling of proteinase inhibitor genes. We believe that complete understanding of this systemic response could have applicability in understanding the broader area of oligosaccharide signalling of defensive genes in plants.

PROTEINASE INHIBITORS AS DEFENSIVE PROTEINS

Proteinase inhibitor proteins are considered to be among the array of defensive chemicals that are present

Table 1. Induction of Various Defensive Responses by Pectic Fragments or by Endopolygalacturonases.

Plant	Response*	Inducer		Ref.**
		PGase	Pectic Fragments	
Ricinus communis	P	+	+	15,16
Glycine max	P	+	+	9,17
Ipomea batatus	P	+	+	18
Trifolium repens	P	---	+	5
Phaseolus vulgaris	P	+	+	5,6
Pisum sativum	P,E,C,G	+	+	19,20
Medicago sativa	PI	---	+	21
Daucus carota	P	+	+	22-24
Capsicun annum	P	+	---	25
Gossypium herbaceum	N,R	+	---	26,27
Cucumis melo	H,E,PI	---	+	28-30
Pinus contorta	P	---	+	31
Cucumis sativus	L	+	+	32
Nicotiana tabacum	H,P,PI	---	+	33
Lycopersicon esculentum	PI	+	+	34
Solanum tuberosum	PI	+	+	35

*Responses: P = Phytoalexin; E = Ethylene; C = Chitinase; G = -Glucanase; N = Necrosis; R = Resistance; H = Hydroxyprotein-rich glycoprotein; PI = Proteinase inhibitor; L = Lignification.

**References

in plants,[36-38] and the biochemistry and molecular biology of their systemically-induced synthesis has been studied in considerable detail.[39,40] Plant proteinase inhibitors were initially isolated from seeds several decades ago, and were studied because of their unusual properties of inhibiting serine protease inhibitors, their nutritional significance in foods (as powerful inhibitors of animal digestive proteases[41,42]), and their usefulness in studying the mechanism of their inhibition of proteases.[30,32] Serine protease inhibitors constitute a significant percentage of the proteins of seeds of important food crops[47,48] and, if not denatured or removed, can be detrimental to the digestive physiology of animals, including man[42] and insects[44] and to the growth of microorganisms.[45] The structure, function and evolution of plant proteinase inhibitors have been the subject of intensive studies for years[36-38] and the plant inhibitors have played a key role in deducing the mechanism of action of the inhibitors in inactivating serine proteinases.[41,43]

In the early 1970s, two proteinase inhibitors, called Inhibitor I, monomer Mr=8100,[46] and Inhibitor II, monomer Mr=12,300,[46] were found to be induced to accumulate in tomato and potato leaves in response to chewing insects.[47,48] A considerable amount of information was already available concerning the structure and properties these two homologous families of inhibitor proteins that greatly facilitated studies of the physiology and molecular biology of their wound-inducibility in leaves.

WOUND-INDUCIBLE PROTEINASE INHIBITORS IN TOMATO AND POTATO LEAVES

Wounding tomato or potato leaves releases into the plant a putative wound signal called the proteinase inhibitor-inducing factor (PIIF).[47,48] This signal is transported out of leaves within 1-2 hr following wounding, and is distributed throughout the plant, but mainly acropetally.[49] Within 4-6 hr following release of the signal (wounding), leaf cells produce newly synthesized Inhibitor I and II mRNA and proteins.[50] Inhibitors I and II mRNAs can be detected in leaves about 2 hr before the inhibitor proteins are detected. The Inhibitor I and II mRNA levels reach a maximum at about 8 hr following wounding and then decline with half-lives of about 10 hr.[50] The synthesis of the inhibitor mRNAs in response to wounding is *de novo* and evidence indicates that they are transcriptionally regulated by the wound signal.[50]

The inhibitor proteins are stored in the central vacuole[51] where they have very long half-lives.[52] Thus, in severely wounded leaves, where mRNA levels reach 0.5% for Inhibitor I and 0.15% for Inhibitor II,[50] the compartmentation and stability of the inhibitors apparently allows them to accumulate while other cellular proteins are made and degraded. Therefore, in response to severe wounding,[50] the dedication of nearly 1% of the leaf mRNA to code specifically for two inhibitors that have long half-lives, allows them to accumulate to high levels in the leaves within a relatively short time.

CELL WALL OLIGOGALACTURONANS INDUCE THE EXPRESSION OF PROTEINASE INHIBITOR GENES

Following the initial observation of the wound-induction of proteinase inhibitors in plants, a major effort spanning several years was made to identify and isolate the proteinase-inhibitor inducing factor, PIIF. Employing an assay system in which small tomato plants were excised with a razor blade and supplied through the cut petioles with extracts from wounded tomato plants to induce proteinase inhibitor synthesis, an active component was identified and purified. A pectic fragment of DP≅30, containing a high percentage of galacturonic acid (Table 1) was identified as the major active component.[14] After partial hydrolysis of the pectic fragment with 2N trifluoric acid,[14] a polygalacturonic acid polymer of DP=20 was produced that was as fully active as its parent cell wall fragment. Enzymic and acid hydrolysis of the active polygalacturonan produced a series of homologous α,1-4 galacturonic acid oligomers from DP=2 to DP=7 that were purified and assayed for PIIF activity.[14] All of the oligomers were active inducers of proteinase inhibitors. The dimer and trimer were about 50% as active as the DP=20 parent (Fig. 1).

Recently, digestion of the DP=20 polygalacturonans with pectin lyase produced unsaturated dimer and unsaturated trimer that were as active as the parent oligomers as inducers of protease inhibitors in tomato plants.[53] These small oligomers have provided new, small chemicals with which to study the proteinase inhibitor inducing mechanisms. The unsaturated dimers and trimers can be labeled with deuterium or tritium by catalytic reduction, and both the saturated and unsaturated dimers and trimers can be reduced with $NaBH_4$ (or NaB^3H_4) to produce an array of derivatives to further investigate the signalling responses. These experiments are presently in progress.[53]

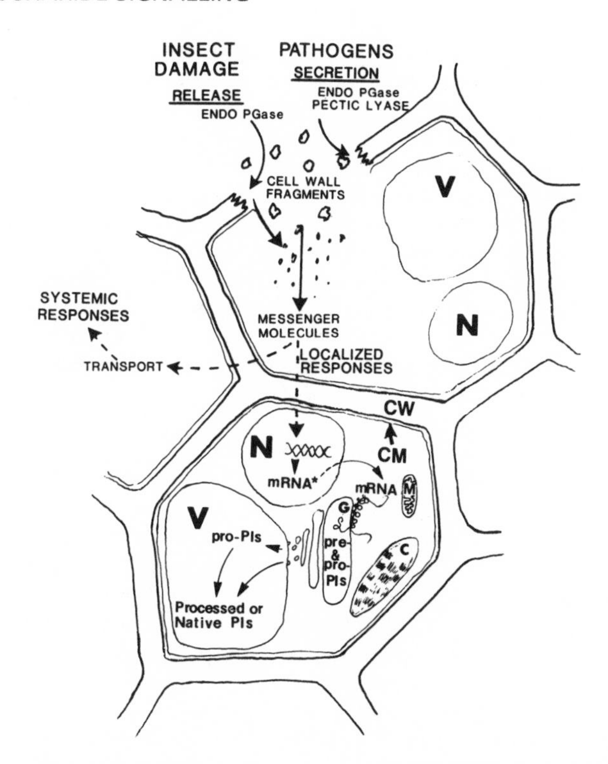

Fig. 2. Postulated sequence of the biochemical events of wound-induced or pathogen-induced synthesis of proteinase inhibitors in plants. The individual pathways of signalling and of proteinase inhibitor mRNA and protein synthesis are discussed in the text.

In Figure 2 is presented an illustration of various aspects of the wound induction that are either known or hypothesized. Several features of the wound induction response have been substantiated: initially, pectic fragments activate inhibitor genes;[50] this activation is transcriptionally regulated and new mRNAs for Inhibitor I and II are synthesized in response to wounding.[50] These messengers then provide for the synthesis of the inhibitors as preproteins[55 56] (Inhibitor I as a preprotein), and the mature proteins are stored in the central vacuole of the leaf.[51] Important aspects of the process still remain obscure. For example, much of the molecular and cell biology of the synthesis and transport of the inhibitors has not been established. It is likely that synthesis takes place in the Golgi if we assume that the process resembles that for deposition of storage proteins (the vacuole-like protein bodies) in developing seeds.[54] However, there is no evidence to support this hypothesis. The vacuole has been likened to the storage bodies and may use the same mechanisms as seeds for synthesis and sorting for vacuolar deposition. On the other hand it is possible that the inhibitors are synthesized on free ribosomes and post-translationally compartmentalized in the vacuole. The resolution of the synthesis and compartmentation of nuclear proteins lies within present methodology and simply remains to be done.

The major questions that this laboratory is seeking to answer are those concerned with the signalling process for the activation of inhibitor genes. It is clear that the addition of the oligogalacturonic acid fragments can activate the genes when supplied to tomato plants through the cut petioles. But how? How are the fragments generated and processed, and are they transported throughout the plants? Are fragments active only in cells surrounding the wound site where they produce a second messenger that is transported throughout the plants? In the first scenario, the fragments would require receptors of some type, either on a membrane or, if internalized, in the cytoplasm - or in both. In the second scenario, a second messenger may be produced, either by a direct biochemical process in the cytoplasm or by the activities of genes that produce a new signal. This system would be inherently more complex. At the present time neither of these possibilities is favored by any of our experiments. In either event the signals are hypothesized to eventually activate or produce trans-acting protein(s) that regulate inhibitor gene expression.

WOUND-INDUCIBLE PROTEINASE INHIBITORS I AND II GENES

The biochemical basis for the signalling response is also being investigated by isolating the genes for the wound-inducible proteinase inhibitors in order to identify the regions where trans-acting factors bind and regulate the wound induction. It is possible that these regions could eventually be used to isolate the trans-acting factors for further studies of the communication system. Wound-inducible Inhibitor I and II genes have been isolated from both the tomato and potato genomes.[56-58] The genes of each family are highly homologous between tomato and potato, exhibiting about 90% identity throughout the open reading frames, including about 50 bp of the 5' region through about 200 bases of the 3' region.

Inhibitor I genes contain two small introns.[55,56] One intron (445 bp in tomato and 479 bp in potato) interrupts the coding region of a signal or transit peptide. The transit peptide of both the tomato and potato Inhibitor I is 23 amino acids in length.[55-58] A pro-sequence is present in both tomato and potato Inhibitor I. This sequence is 19 amino acids in length in the tomato genome, but contains only 13 residues in the potato gene prosequence.

The native Inhibitor I found in tomato leaves is 69 amino acids in length[55 59] and in potato leaves is 71 amino acids in length.[58] The difference in length between potato and tomato Inhibitor I proteins is a result of a difference in processing sites.

The wound-inducible Inhibitor II gene from both potato and tomato contains a single intervening sequence of 117 bp found within the transit sequence of the protein.[56,58] The transit sequence is 30 amino acids in length and, when processed, the native inhibitor contains 123 residues. The protein is clearly gene duplicated and elongated and possesses two inhibitory reactive sites.[56,59,60]

Inhibitor I and II are not related and represent two of the six known plant proteinase inhibitor families (Table II).[43] No homology is evident between either their amino acid sequences or their nucleotide sequences. Only short nucleotide sequences at putative regulatory regions are similar between the two inhibitors, i.e., TATA boxes, polyadenylation signals and poly A-rich regions just 5' to the translation initiation. One small region is common to both Inhibitor I and II genes: a palindromic sequence of CATTATAATG about 40 bp 5' to the polyadenylation signal.[55-60] A role for this sequence has not yet been established, but recent experiments in which the 5' and 3' regions of the Inhibitor II gene were used to transform

Table 2. Families of Protein Inhibitors of Serine Proteinases*

	Animals
1.	Bovine pancreatic trypsin inhibitor (Kunitz) family
2.	Pancreatic secretory trypsin inhibitor (Kazal) family
3.	*Ascaris* inhibitor family
4.	Serpin family (mechanistically distinct)
5.	Hirudin family
	Plants
6.	Soybean trypsin inhibitor (Kunitz) family
7.	Soybean proteinase inhibitor (Bowman-Birk) family
8.	Potato 1 family
9.	Potato 2 family
10.	Barley trypsin inhibitor family
11.	Squash inhibitor family
	Microbial
12.	*Streptomyces subtilisin* inhibitor (SSI) family
13.	OTHER FAMILIES

*M. Laskowski, Jr. *et al*. Cold Spring Harbor Symposium in Quantitative Biology (in press).

tobacco plants with an inducible chloramphenicol acetyl transferase (CAT) gene indicate that an element in the 3' region may be partly responsible for wound induction. The possible involvement of this small palindrone in the wound inducibility of the inhibitors is currently under investigation.

REGULATION OF A WOUND-INDUCIBLE PROMOTER IN TRANSGENIC PLANTS

A transcriptional fusion comprised of about 1 kb of the potato 5' region Inhibitor II gene was constructed with a chloramphenicol-acetyl transferase gene at a restriction site 18 bp 5' from the Inhibitor II translation initiation codon.[59] This Inhibitor II 5'-CAT gene fusion was terminated with 1 kb of the Inhibitor II 3' terminator that included 11 bp of the C-terminal region of the Inhibitor II open reading frame, and with a constitutive terminator from transcript 6b of the Ti plasmid. These two chimeric genes were transferred to tobacco cells using a binary Ti vector system and the resulting transgenic plants were regenerated.[59] The CAT gene was wound-inducible in leaves of tobacco plants having both the Inhibitor IIK promoter and terminator, but not in plants with the transcript 6b terminator.[57] The CAT gene was expressed both locally (in the same leaf near the wounds) and systemically (in adjacent, upper leaves). Thus, although tobacco leaves do not express wound-inducible Inhibitor I or II in response to wounding, they possess the proper factors to regulate the expression of the CAT gene under control of the potato wound-inducible Inhibitor II promoter and terminator.

The lack of response of the chimeric gene lacking the Inhibitor IIK terminator suggests that an element in the 3' region is responsible, at least in part, for the wound-inducibility of the CAT gene.[57] The 5' and 3' regions of the Inhibitor II genes are currently under investigation to identify the specific sequences responsible for wound-inducibility. When such regions are identified, they can be employed to seek the transacting factors that are responsible for the wound-induction. It is hoped that such proteins may bring us closer to the understanding of how oligosaccharide fragments, released from the cell wall during wounding, can activate defensive genes in plants.

SUMMARY

The property of oligogalacturonan fragments, derived from the plant cell wall, to activate a variety of localized and systemic defensive responses in plants indicates that an intracellular signalling system may be present throughout the plant kingdom to regulate the expression of defensive genes in response to pest attacks. The induction of proteinase inhibitor genes by small oligouronides has been studied as a peridigm to unravel the mechanisms of oligosaccharide signalling. Di- and trigalacturonides have been isolated that initiate proteinase inhibitor synthesis in leaves of tomato plants and now provide small probes to further study the mechanisms of their reception in plants. Wound-inducible proteinase inhibitor genes have been isolated and characterized. The 5' and 3' regulatory regions of a potato Inhibitor IIK gene have been fused with a reporter gene (chloramphenicol acetyl transferase, CAT) and successfully used to transform tobacco plants with a wound-inducible CAT gene. The cis-acting sequences and the transacting factors that regulate the wound expression are currently being sought for use in identifying and isolating proteins that regulate the wound-inducible expression of the proteinase Inhibitor II gene. It is anticipated that these approaches will facilitate understanding of the biochemical basis for oligosaccharide signalling of genes involved in plant defensive responses.

ACKNOWLEDGMENTS

The author wishes to acknowledge the efforts of all of the students, postdoctoral fellows, technicians and visiting scientists that contributed to the research from my laboratory. Also acknowledged is support from Washington State University, The National Science Foundation and the U.S. Department of Agriculture.

REFERENCES

1. RHODES, D.F. 1979. Evaluation of plant chemical defense against herbivores. In Herbivores, Their Interaction With Secondary Plant Metabolites. (G.A. Rosenthal, D.H. Janzen, eds.), Academic Press, New York, pp. 1-55.
2. KOLATTUKUDY, P.E. 1980. Biopolyester membranes of plants, cutin and suberin. Science 208: 990-1000.
3. RHODES, D.F. 1983. Herbivore population dynamics and plant chemistry. In Natural and Managed Systems. (R.F. Denno, M.S. McClare, eds.), Academic Press, New York, pp. 155-220.

4. ALBERSHEIM, P., A.G. DARVILL, M. McNEIL, B. VALENT, J.K. SHARP, E.A. NOTHNAGEL, K.R. DAVIS, N. YAMAZAKI, D.J. GOLLIN, W.S. YORK, W.F. DUDMAN, J.E. DARVILL, A. DELL. 1983. Oligosaccharins, naturally occurring carbohydrates with biological regulatory functions. In Structure and Function of Plant Genomes. (O. Ciferri, L. Dure, III, eds.), Plenum Press, New York, pp. 293-312.
5. WEST, C.A., R.J. BRUCE, D.F. JIN. 1984. Pectic fragments of plant cell walls as mediators of stress responses. In Structure, Function, and Biosynthesis of Plant Cell Walls. (W.M. Dugger, S. Bartnicki-Garcia, eds.), Waverly Press, Baltimore, pp. 359-380.
6. RYAN, C.A., P.D. BISHOP, J.S. GRAHAM, R.M. BROADWAY, S.S. DUFFEY. 1986. Plant and cell wall fragments activate proteinase inhibitor genes for plant defense. J. Chem. Ecol. 12: 1025-1035.
7. AYRES, A., B. VALENT, J. EBEL, P. ALBERSHEIM. 1976. Host-pathogen interactions. XI. Composition and structure of wall-released elicitor fractions. Plant Physiol. 57: 766-744.
8. WEST, C.A., P. MOESTA, D.F. JIN, A.F. LOIS, K.A. WICKHAM. 1985. The role of pectic fragments of the plant cell wall in the response to biological stresses. In Cellular and Molecular Biology of Plant Stress. (J.L. Key, T. Kosuge, eds.), Alan R. Liss, Inc., New York, pp. 335-350.
9. HAHN, M.G., A.G. DARVILL, P. ALBERSHEIM. 1981. Host-pathogen interactions: XIX. The endogenous elicitor, a fragment of a plant cell wall polysaccharide that elicits phytoalexin accumulation in soybeans. Plant Physiol. 68: 1161-1169.
10. BISHOP, P., D.J. MAKUS, G. PEARCE, C.A. RYAN. 1981. Proteinase inhibitor inducing factor activity in tomato leaves resides in oligosaccharides enzymically released from cell walls. Proc. Natl. Acad. Sci. USA 78: 3536-3640.
11. RYAN, C.A., P. BISHOP, G. PEARCE, A.G. DARVILL, M. McNEIL, P. ALBERSHEIM. 1981. A sycamore cell wall polysaccharide and a chemically related tomato leaf polysaccharide possess similar proteinase inhibitor-inducing activities. Plant Physiol. 68: 616-618.
12. NOTHNAGEL, E.A., M. McNEIL, P. ALBERSHEIM, A. DELL. 1983. Host-pathogen interactions: XXII. A

galacturonic acid oligosaccharide from plant cell walls elicits phytoalexins. Plant Physiol. 71: 916-926.

13. JIN, D.J., C.A. WEST. 1984. Characteristics of galacturonic acid oligomers as elicitors of casbene synthetase activity in castor bean seedlings. Plant Physiol. 74: 989-992.
14. BISHOP, P., G. PEARCE, J.E. BRYANT, C.A. RYAN. 1984. Isolation and characterization of the proteinase inhibitor inducing factor from tomato leaves: identity and activity of poly- and oligogalacturonide fragments. J. Biol. Chem. 259: 13172-13177.
15. LEE, S.-C., C.A. WEST. 1981. Properties of *Rhizopus stolonifer* polygalacturonase, an elicitor of casbene synthetase activity in castor bean (*Ricinus communis* L.) seedlings. Plant Physiol. 67: 640-645.
16. BRUCE R.J., C.A. WEST. 1982. Elicitation of casbene synthetase activity in castor bean: the role of pectic fragments of the plant cell wall in elicitation by a fungal endopolygalacturonase. Plant Physiol. 69: 1181-1188.
17. DAVIS, K.R., A.G. DARVILL, P. ALBERSHEIM. 1986. Host-pathogen interactions: XXX. Characterization of elicitors of phytoalexin accumulation in soybean released from soybean cell walls by endopolygalacturonic acid lyase. Z. Naturforsch. 41c: 39-48.
18. SATO, D., I. URITANI, T. SAITO. 1982. Properties of terpene-inducing factor extracted from adults of the sweet potato weevil, *Cylas formicarius* Fabricius (Coleoptera:Brenthidae). Appl. Entomol. Zool. 17: 386-374.
19. WALKER-SIMMONS, M., L. HADWIGER, C.A. RYAN. 1983. Chitosans and pectic polysaccharides both induce accumulation of the antifungal phytoalexin pisitin in pea pods and antinutrient proteinase inhibitors in tomato leaves. Biochem. Biophys. Res. Commun. 110: 194-199.
20. WALKER-SIMMONS, M., D. JIN, C.A. WEST, L. HADWIGER, C.A. RYAN. 1984. Comparison of proteinase inhibitor-inducing activities and phytoalexin elicitor activities of a pure fungal endopolygalacturonase, pectic fragments, and chitosans. Plant Physiol. 76: 833-836.
21. BROWN, W., C.A. RYAN. 1984. Isolation and characterization of a wound-induced trypsin inhibitor

from alfalfa leaves. Biochemistry 23: 3418-3422.
22. KUROSAKI, F., K. FUTAMURA, A. NISHI. 1985. Fractors affecting phytoalexin production in cultured carrot cells. Plant Cell Physiol. 26: 693-700.
23. KUROSAKI, F., A. NISHI. 1984. Elicitation of phytoalexin production in cultured carrot cells. Physiol. Plant Pathol. 24: 169-176.
24. KUROSAKI, F., Y. TSURUSAWA, A. NISHI. 1985. Partial purification and characterization of elicitors for 6-methoxymellein production in cultured carrot cells. Physiol. Plant Pathol. 27: 209-217.
25. WATSON, D.G., C.J.W. BROOKS. 1984. Formation of caposidol in Capsicum annum fruits in response to non-specific elicitors. Physiol. Plant Pathol. 24: 331-337.
26. KARBAN, R. 1985. Resistance against spider mites in cotton induced by mechanical abrasion. Entomol. Exp. Appl. 37: 137-141.
27. VENERE R.J., L.A. BRINKERHOFF, R.K. GHOLSON. 1984. Pectic enzyme: an elicitor of necrosis in cotton inoculated with bacteria. Proc. Okla. Acad. Sci. 64: 107.
28. ROBY, D., A. TOPPAN, M.T. ESQUERRE-TUGAYE. 1985. Cell surfaces in plant-microorganisms interactions. V. Elicitors of fungal and of plant origin trigger the synthesis of ethylene and of cell wall hydroxyproline-rich glycoproteins in plants. Plant Physiol. 77: 700-704.
29. ROBY, D., A. TOPPAN, M.T. ESQUERRE-TUGAYE. 1986. Cell surfaces in plant-microorganism interactions. VI. Elicitors of ethylene from Colletotrichum lagenarium trigger chitinase activity in melon plants. Plant Physiol. 81: 228-233.
30. ESQUERRE-TUGAYE, M.T., D. MAZAU, B. PELISSIER, D. ROBY, D. RUMEAU, A. TOPPAN. 1985. Induction by elicitors and ethylene of proteins associated to the defense of plants. In J.L. Key, T. Kosuge, eds., op. cit. Reference 8, pp. 459-473.
31. MILLER, R.H., A.A. BERRYMAN, C.A. RYAN. 1986. Biotic elicitors of defense reactions in lodgepole pine. Phytochemistry 25: 611-612.
32. ROBERTSEN, B. 1986. Do galacturonic acid oligo-saccahrides have a role in the resistance mechanism of cucumber towards Cladosporium cucumerinum? In Biology and Molecular Biology of Plant-Pathogen

Interactions. (J. Bailey, ed.), Springer-Verlag, Berlin, pp. 177-183.
33. BAKER, G.S., M. ATKINSON, A. COLLMER. 1985. Effects of cell wall fragments released by pectate lyase on the hypersensitive response in tobacco. Phytopathology 75: 1373.
34. MODDERMAN, P.E., C.P. SCHOT, F.M. KLIS, D.H. WIERINGA-BRANTS. 1985. Acquired resistance in hypersensitive tobacco against tobacco mosaic virus, induced by plant cell wall components. Phytopathol. Z. 113: 165-170.
35. WALKER-SIMMONS, M., C.A. RYAN. 1977. Wound-induced accumulation of trypsin inhibitor activities in plant leaves: a survey of several plant genera. Plant Physiol. 59: 437-439.
36. RYAN, C.A. 1973. Proteolytic enzymes and their inhibitors in plants. Annu. Rev. Plant Physiol. 24: 173-196.
37. RICHARDSON, M. 1977. The proteinase inhibitors of plants and microorganisms. Phytochemistry 16: 159-169.
38. RYAN, C.A. 1979. Proteinase inhibitors. In The Biochemistry of Plants. (P.K. Stumpf, E.E. Conn, eds.), Academic Press, Inc., New York, Vol. 6, pp. 351-371.
39. RYAN, C.A. 1984. Systemic responses to wounding. In Plant Microbe Interactions: Molecular and Genetic Perspective. (T. Kosuge, E.W. Nester, eds.), MacMillan Publishing Co., Vol. 1, pp. 307-320.
40. RYAN, C.A., P. BISHOP, G. PEARCE, M. WALKER-SIMMONS, W. BROWN. 1984. Pectic fragments regulate the expression of proteinase inhibitor genes in plants, UCLA Symposium. In J.L. Key, T. Kosuge, eds., op. cit. Reference 8, pp. 319-334.
41. LASKOWSKI, M., JR., R.W. SEALOCK. 1977. Protein proteinase inhibitors: molecular aspects. In The Enzymes. (P.D. Boyer, ed.), Academic Press, New York, Vol. 3, pp. 375-473.
42. LIENER, I.E., M.L. KAKADE. 1969. Protease inhibitors. In Toxic constitutents of Plant Foodstuffs, 2nd Ed. (I.E. Liener, ed.), Academic Press, pp. 8-66.
43. LASKOWSKI, J., JR., I. KATO. 1980. Protein inhibitors of proteinases. Annu. Rev. Biochem. 49: 593-626.
44. BROADWAY, R.M., S. DUFFEY. 1986. Plant proteinase inhibitors: mechanism of action and effect on the growth and digestive physiology of larval *Heliothies*

zea and Spodoptera eniqua. J. Insect Physiol. 32: 827-833.

45. SENSER, F., H.-D. BELITZ, K.-P. KAISER, K. SANTARIUS. 1974. Suggestion of a protective function of proteinase inhibitors in potato: inhibition of proteolytic activity of microorganisms isolated from spoiled potatoes. Z. Lebensm. Unters-Forsch 155: 100-101.
46. PLUNKETT, G., D.F. SENEAR, G. ZUROSKE, C.A. RYAN. 1982. Proteinase Inhibitor I and II from leaves of wounded tomato plants: purification and properties. Arch. Biochem. Biophys. 213: 463-472.
47. GREEN, T.R., C.A. RYAN. 1972. Wound-induced proteinase inhibitor in plant leaves: a possible defense mechanism against insects. Science 175: 776-777.
48. RYAN, C.A. 1978. Proteinase inhibitors in plant leaves: a biochemical model for pest-induced natural plant protection. Trends Biochem. Sci. 5: 148.
49. NELSON, C.E., M. WALKER-SIMMONS, P. MAKUS, G. ZUROSKI, J. GRAHAM, C.A. RYAN. 1983. Regulation of synthesis and accumulation of proteinase inhibitors in leaves of wounded tomato plants. In Plant Resistance to Insects. (P. Hedin, ed.), American Chemical Society Press, Washington, D.C., pp. 108-122.
50. GRAHAM, J.S., G. HALL, G. PEARCE, C.A. RYAN. 1986. Regulation of synthesis of proteinase Inhibitors I and II mRNAs in leaves of wounded tomato plants. Planta 169: 399-405.
51. WALKER-SIMMONS, M., C.A. RYAN. 1977. Immunological identification of proteinase Inhibitors I and II in isolated tomato leaf vacuoles. Plant Physiol. 60: 61-63.
52. GUSTAFSON, G., C.A. RYAN. 1976. The specificity of protein turnover in tomato leaves: the accumulation of proteinase inhibitors, induced with the wound hormone PIIF. J. Biol. Chem. 251: 7004-7010.
53. PEARCE, G., T.M. MOLOSHOK, C.A. RYAN. 1988. (In preparation).
54. CHRISPEELS, M. The role of the golgi apparatus in the transport and post-translational modification of vacuolar (protein body) proteins. Oxf. Surv. Plant Mol. Biol. 2: 43-68.

55. GRAHAM, J., G. PEARCE, J. MERRYWEATHER, K. TITANI, L. ERICSSON, C.A. RYAN. 1985. Wound-induced proteinase inhibitor mRNA from tomato leaves: I. The cDNA-deduced sequence of pre-Inhibitor I and its post-translational processing. J. Biol. Chem. 260: 6555-6560.
56. GRAHAM, J.S., G. PEARCE, J. MERRYWEATHER, K. TITANI, L.H. ERICSSON, C.A. RYAN. 1985. Wound-induced proteinase inhibitor mRNA from tomato leaves: II. The cDNA-deduced primary sequence of pre-Inhibitor II. J. Biol. Chem. 260: 6561-6564.
57. THORNBURG, R.W., T.E. CLEVELAND, C.A. RYAN. 1988 Wound-inducible expression of potato Inhibitor II gene in transgenic tobacco plants. Proc. Natl. Acad. Sci. USA (in press).
58. CLEVELAND, T.E., R.W. THORNBURG, C.A. RYAN. 1987. Molecular characterization of a wound-inducible proteinase inhibitor gene from potato and the processing of the mRNA and protein. Plant Mol. Biol. 8: 199-207.
59. LEE, J.S., W.E. BROWN, G. PEARCE, T.W. DREHER, J.S. GRAHAM, K.G. AHERN, G.D. PEARSON, C.A. RYAN. 1986. Molecular characterization and phylogenetic studies with a wound-inducible proteinase inhibitor gene in Lycopersicum species. Proc. Natl. Acad. Sci. USA 83: 7277-7281.
60. FOX, E., C.A. RYAN. 1988. (In preparation).

Chapter Nine

PHYTOCHEMISTRY -- ITS ROLE IN THE FUTURE OF PLANT BIOTECHNOLOGY 1987

LEON DURE III

Department of Biochemistry
University of Georgia
Athens, Georgia 30602

INTRODUCTION

Knowing full well that I know nothing whatsoever about secondary metabolism and less about fermentation, Dr. Kosuge has asked me to give my perspective as a biochemist of plant biotechnology. In view of these limitations, my perspective must be confined to that area of plant biotechnology that I try to keep up with, that based upon recombinant DNA technology.

THE MYTH OF SCARCITY

Very few farmers seemed to ever make much money when I was a young boy growing up on a farm in Virginia. We all worked hard and lived a first-rate life, but the financial returns never seemed to me to match the effort. However, recognizing the population growth and the diminishing number of farmers, I felt that it was only a matter of about 5 years before farmers, a rare breed furnishing the most fundamental of needs, would be able to call the financial tune. This was about 1950. In 5 years nothing seemed to have changed - nor in 5 more years - nor in the 5 after that. Now in 1987 (37 years later) there are even less farmers, but still they work very hard for little return. The factor I failed to consider was that the increase in demand was to be matched by an increase in yield: not just in the United States but also in Europe and most of Asia. This increase has been so dramatic that twice efforts were made in our country to keep farmers from farming so much, i.e. the Soil Bank of the late 50's and the recent Pic Program.

The yield increase has been the result of a number of factors such as irrigation, increased use of fertilizer, and more land in production. However, the major cause has probably been the success of plant breeding. It is hard to imagine, but since 1960 there has been almost a 3-fold increase in rice yield per hectare in Asia and in wheat yield per hectare in Europe. A truly startling figure is that the International Rice Research Institute in Manila acquired 90,000 different strains of rice in helping to bring about the green revolution in rice.

To keep farmers just a little mollified, price supports were established in the United States, Europe and Japan and these have falsified the supply and demand picture ever since. Worldwide, agricultural subsidies will amount to ~\$150 billion in 1987.

Surely there are areas of the world where people barely get enough to eat and, in periods of local poor harvests, fail to get enough. However, this is not because the world does not, nor cannot, produce enough calories and proteins and vitamins, but because these regions have so little to sell their fellowman that they have insufficient funds to buy from the world's surpluses.

In fact, I have heard that, were farming practices universally efficient, over 10 billion souls could be susptained on our earth. In view of this one could make a case for a moratorium on agricultural research in the developed world.

So what really keeps the price down and farmers unprosperous is that there are no overall shortages;

instead there are overall surpluses. People are hungry but not because there is a lack of food.

What does all this have to do with biotechnology? Simply that at the moment world hunger cannot be the critical driving force to do recombinant DNA work in plants on an applied level. And thank goodness it is not, for in my opinion we do not know enough about how plants work to meet the challenge.

EARLY ENTHUSIASMS HAVE WILTED

The recombinant DNA age began in earnest about 1980 when it was realized that, by moving genes around, mankind had the capability of finally, as Descartes wished, "becoming the masters and possessors of nature", at least of biological nature. In the plant world the specter of millions of people that were hungry most of the time caused most people to feel that there was then, and perhaps always had been, a shortage of food worldwide. Now this was indeed a challenge for recombinant DNA. Given a few years, plants would be so modified as to make the desert bloom, provide seeds that were nutritionally perfect, and plants that insects, fungi and bacteria could not stomach. In short, food scarcity would become something of the past.

The reaction of the medical world to these new possibilities were rather more specific. The medical doctors and research scientists wanted specific things from recombinant DNA for specific reasons: non-antigenic insulin, human growth hormone, vaccines for hepatitis B and calf scours, and a myriad of simple, reliable diagnostics. The wish list of the plant world was much more fuzzy and vague: pest resistance, salt and drought tolerance, improved photosynthetic and nitrogen fixation efficiencies, nutritionally complete seeds and, as always, increased yields. In short, better plants without a detailed knowledge of what we meant by better.

To bring forth the harvest based on recombinant DNA an astounding number of small biotechnology companies were formed. There was no shortage of investors willing to gamble on this technology nor of courageous scientists who were willing to give up tenure, security and the comfortable, assured life of universities to run these companies. A surprising number of these companies were exclusively devoted to designing better or tailor-made plants. Thank goodness for the plant companies because, as I will point out later, neither botany departments nor the USDA were really aware of what was going on.

Today we recognize the naivete in assuming that there was worldwide scarcity that would serve as the impetus for research and discovery in recombinant DNA as applied to plants. Today I think we also recognize that in the early days of our enthusiasm we failed to realize that we did not know how to design a better plant. It was easy to say that recombinant DNA would allow us to feed an exploding worldwide population and use less fertilizer, energy and drudgery in the process. It was impossible to say precisely how this was to be accomplished. No one bothered to ask "what genes should we move?". We now realize this. Stock in the private ventures is no longer sold to put nitrogen fixation into wheat, or to farm the desert or tundra, or to change the composition of seed proteins. We now realize that to put nitrogenase and its ancillary proteins into wheat would require enormous engineering of the wheat genome and would probably starve the wheat plant for energy when achieved. As for farming inhospitable terrain we now recognize that crossing in unknown genes from non-economic relatives of the plant that happen to be adapted to the environment is a much easier route. In other words, this may be possible to achieve by classical plant breeding. After all, natural selection has been working towards such adaptations for 150 million years. As for changing the nutritional aspects of seeds, we found that we must deal with many, many genes. The globulins of dicots probably emanate from 15 to 30 genes and the prolamines of cereals from over 100. In short, the menu of possible targets for the genetic engineering of plants has become much more constrained. Put another way, the "bloom is off the rose".

There are of course attractive things for private ventures to do at present. Herbicide resistance is being engineered through recombinant DNA technology and may cut farmer's costs, thought it probably won't decrease the use of agrichemicals. Predator resistance seems to have focused on efforts to express the various "b.t. proteins" of *Bacillus thuringenesis* in leaves. New methods for transforming plants are being established serendipitously.

TWENTY-FIVE YEARS OF RESEARCH DOLDRUMS

It may be asked, "why was the plant world unprepared for the revolution in recombinant DNA?". Why was the plant wish list so fuzzy and naive? The answer to that is simply that the plant world did not participate to a large extent in the biochemical revolution that made the recombinant DNA revolution possible. From 1950 to 1975 there were not many individuals who made a living opening up plant cells and asking what is going on inside.

The USDA never quite understood that what occurs on the gross, visible level is the result of what is occurring on the invisible level on the inside. There was always hybrid vigor to serve as the conceptual background (first described in 1904) and a lot of spraying and fertilizing to be done, and furthermore, yields continued to rise. Botany departments on the other hand fought research on the molecular level with tooth and nail. No hot beds of plant biochemistry were allowed to develop. The few individuals who did carry out plant biochemistry were never part of a critical mass that fed upon each other's findings. And progress was slow. I imagine that in 1960 there were 10 times the number of individuals studying metabolism in liver than in all plants combined.

The paucity of effort and results made biochemistry a preserve of animal and bacterial science. Teachers of biochemistry would say that since certain processes take place in *Escherichia coli* and in rat liver, they must take place in plants also, although no one had bothered to look. Most text books treated plant cells as animal cells with chloroplasts encased in wood! Now the basic unity of intermediary metabolism throughout the biological world is a marvelous thing and simply points out that the compounds and the syntheses hit upon by the organism that gave rise to all eucaryotes have never been improved upon. However, as we know, plants are not just stationary animals, and a knowledge of animal processes cannot help us much longer. Having missed out on the first biochemical revolution, we now must start one of our own based on the unique features of plants - the plant biochemical revolution.

BLACK BOXES

As we know, the Malthusian prediction cannot be delayed forever. Population will someday out pace yield, a yield based on what we know now. Since we do not have to do battle with immediate scarcity, we have a respite in time. A respite of perhaps 30 years before we actually must have super plants, tailor-made plants, plants embodying the wish list. My feeling is that we must use this time to establish a reservoir of basic biochemical information, to find out how plants really work so that we will know what we need to modify, enhance, or inhibit in order to develop the super plants through genetic engineering.

To do this we need to establish the critical masses and to recognize our identity as plant biochemists and the uniqueness of plant biochemistry. Plant intermediary metabolism and its regulation must be distinct from that of bacteria and higher animals, and this should be reflected in a plant's complement of enzymes and their regulation. Being totipotent nutritionally, plants do not have to balance the intake, degradation, or excretion of organic molecules. A plant's diet is CO_2, water, light and some mineral salts. Further, since plants do not fight or run, their metabolic regulation cannot be based on this choice, but rather on the monitoring of physical entities such as water, light and temperature. Finally, plants have their own set of pathogens and stresses as well as their own mechanisms for coping with them.

There are signs that plant biochemistry is beginning to exploit the unique features of plants, and marvelous experimental observations are becoming numerous. However, there are areas that I personally see as black boxes that need to be opened up and explored. Let me list a few.

Cell Surface Receptors

Animals have sophisticated sets of extra cellular signal molecules and their cell surface receptors that impose direction on metabolism from outside the cell. Some of these direct morphogenesis, others direct energy metabolism. Intricate circuits involving morphogenic transducers and protein kinases work in animal cells; likewise, Ca^{++}, cyclic AMP, cyclic GMP and inositol phosphates perform roles in a cascade of interactions that govern homeostasis as well as reactions to stimuli.

Do plants have an equivalent system? What information do plant cells exchange? Is it all done by plasmodesmatal interchange, or are there receptors for informational molecules travelling in the phloem? Evidence for a few cell surface hormone receptors has been gathered. Small oligosaccharides travelling in the phloem are known to elicit response on the gene level in the case of proteinase inhibitor synthesis, in the wake of insect damage, and perhaps in the transmission of the flowering stimulus. These molecules must have receptors and elicit cascades of reactions in recipient cells. Let's find them.

Integration of Intermediary Metabolism

We know from the biochemistry of animals that rarely are enzymes a limiting factor in metabolism. They are rarely even half-saturated. Metabolism is substrate driven and the directions of and the rate of flux of metabolites is allosterically controlled. The concentration of some few enzymes, however, is critical and plays a decisive role in metabolism. Another truism in metabolism is that pathways do not operate *in vacuo*. A perturbation in one pathway probably has ramifications throughout intermediary metabolism. If we don't know how these pathways are integrated and dominated by central carbohydrate metabolism, how are we to ever modify a plant's characteristics in a meaningful fashion?

A case can be made that plants regulate metabolism solely by the most fundamental and primitive of all mechanisms, i.e. mass action. Since they may not need a repertoire of fast responses and do not need to balance intake and synthesis with degradation, mass action and a little feedback allosterism may be all that is needed. I don't think it will be that simple. After all, plants have cyclic AMP and calmodulin, but we have uncovered very little use for either.

The integration of metabolism can be studied in many forms. To me, the most fascinating is that which coordinates the activities of the cytosol and the plastids. We are just beginning to see how a plant cell decides between storing carbon as starch or exporting carbon as sucrose. The involvement of fructose-2,6-bisphosphate in regulating glycolysis and of 2-carboxyarabitol phosphate as a nocturnal inhibitor of ribulose-1,5-bisphosphate carboxylase-oxygenase adds to the complexity of the regulation of this decision.

We know very little about how fatty acid or amino acid synthesis is integrated between the two compartments, although it is realized that most of the steps take place in the plastid utilizing procaryotic-like enzymes but that some steps are cytosolic. How is movement of precursors and products through plastid membrane handled? When we have a composite picture of the overall regulation of these processes, we will have something to cheer about, indeed.

Amino acid metabolism by itself has some black areas in my view. We know most of the steps in most of the pathways which differ only trivially from those of bacteria. Yet, other than a few examples from the aspartic acid family, we are mostly ignorant of the points and extent of allosteric regulation. The mammalian degradation of amino acids for energy is well worked out. Every student knows that he or she cannot live on lysine or leucine as carbohydrate sources, since they only yield acetate. Do plants degrade amino acids at all, other than to generate precursors for other end products? Is there a need for plants to do so? Is the regulation of synthesis enough to insure a proper control over their concentration?

Further, how is the synthesis of amino acids regulated by carbohydrate metabolism? It must be. The flow of carbon for energy generating metabolism must have priority. A nice example of this in all organisms is the thermodynamic barrier between glucose-6-phosphate and glucose-1-phosphate. By this simple means, nature insures that demands for ATP and NADH are satisfied before carbon is stored or polymerized.

The fact that two enzymes in amino acid synthesis are the targets of two of the major herbicide groups should make amino acid metabolism popular. These are the inhibition of acetolactate synthase by phenylureas and of enolpyruvyl shikimate phosphate synthase by glyphosate.

Cell Wall Synthesis

We have yet to have an efficient cell-free system for cellulose synthesis. The enzyme complex cellulose synthetase seems to need membranes, or perhaps to be embedded in membranes in order to catalyze the condensations. The biophysics of the interactions of cellulose, arabinoglycans, pectins and cell wall proteins that produce the great variety of cell walls is in its infancy. What we are finding out chemically about the hydroxyproline-rich and glycine-rich cell wall proteins is opening new vistas to these structures.

MOBILIZATION

What I have outlined is a lot of work, and you may ask, "who is going to do it?". Most graduate students want to manipulate genes. They are terrified of proteins. They turn up their noses at being enzymologists or protein chemists, or at trying to understand how the biosynthesis of compound X is regulated. How can we change this? Principally by our own enthusiasm for what we do. Progress generates excitement and excitement is contagious. We are making progress as we approach a critical mass in some areas, and we must transmit this excitement. It does not hurt to point out that DNA does nothing. Proteins do all the work. And if we don't understand how the proteins work, we really don't comprehend the process. In fact, all the advances in recombinant DNA technology stem from advances in enzymology. What can a cloner do without the ligases, polymerases, nucleases, phosphatases and restriction enzymes that they purchase by mail. In the future it will be up to us to tell the genetic engineers what to engineer. If we point out, with a little élan, the fascinating biochemical problems unique to plants waiting to be solved, the students will appear.

Secondly, we must throw our weight around in our departments and schools. We should insist that courses in general biochemistry be rid of their medical bias and contain more plant biochemistry than just photosynthesis.

After all, students want to do what they have heard about. In many institutions, what we do is a secret. Further, we must establish more courses in plant biochemistry that assume a knowledge of general biochemistry. We need to spend the time on the unique features of plant metabolism at their forefronts. A good one semester graduate level text would help here. Do you know of a text book that points out why it is a marvelous temporary response for plants experiencing anaerobic stress to synthesize lots of proline? Proline is a first-rate, innocuous, anaerobic reservoir for hydride ions that allows NAD^+ to cycle and glycolysis to continue. And when the O_2 stress is relieved, the high energy hydride ions are recovered, since its synthesis is reversible. Such tidbits turn graduate students on to the subject.

Further, we must lobby for more positions for plant biochemists so that we do not work *in vacuo*, but profit from daily contact with different aims and differing points of view.

Lastly, we must demand that the USDA finally become the NIH of plant basic research. This should take the form of a much expanded competitive grants program that gives many more awards and much larger sums, and in the form of institutional training grants. This latter program would serve not only to give our work a sense of identity, but would emphasize to deans that clusters of plant biochemists are a fashionable thing.

In lobbying for all this, we should justify it, not as something having an economic payoff, but as basic science has always been justified -- as a fruitful intellectual endeavor, essential for human progress, but driven by human curiosity. Feeding the world will come as a natural consequence.

INDEX